Viewdata in Action

Viewdata in Action
A Comparative Study of Prestel

Editor: **Rex Winsbury**

McGRAW-HILL Book Company (UK) Limited

London ·	New York ·	St Louis ·	San Francisco ·	Auckland
Bogota ·	Guatemala ·	Hamburg ·	Johannesburg · Lisbon ·	Madrid
Mexico ·	Montreal ·	New Delhi ·	Panama ·	Paris · San Juan
São Paulo ·	Singapore ·	Sydney ·	Tokyo ·	Toronto

Published by
McGRAW-HILL Book Company (UK) Limited
MAIDENHEAD · BERKSHIRE · ENGLAND

British Library Cataloguing in Publication Data

Viewdata in Action.
  1. Viewdata (Data transmission system)
  I. Winsbury, Rex
  384.55′4   TK5105   80–41262

ISBN 0-07-084548-4

12345   M & G   83210

Printed and bound in Great Britain by
Morrison & Gibb Ltd, London and Edinburgh

Contents

Discrimination between the sexes

In a contributed book of this type, it is impracticable to eliminate completely the use of genders and retain a fluent text. The reader is asked to accept that any deliberate distinction between the sexes is not implied.

Contributors

Stephen Castell, with a background in industrial research, computer consultancy and merchant banking, in November 1978 set up Castell Computer & Systems Telecommunications Ltd as an independent management and financial consultancy in information technology. He has been instrumental in bringing some significant and novel business services to Prestel, notably Infolex, the UK's first commercially-available legal information retrieval service for practising lawyers, and BBC Data (to which he is appointed consultant), the British Broadcasting Corporation's new business-information service.

T. F. Chapman is Viewdata Services Manager with Baric Computing Services Ltd, a major UK computer bureau owned by International Computers Ltd and Barclays Bank. He was responsible for his own companies' participation in viewdata in 1977. Having joined Baric as Market Development Manager in 1975, he was previously with ICL involved with corporate planning and business modelling and marketing. During his period of involvement in the viewdata service industry, Baric has emerged as the leading umbrella information provider on Prestel and has played a major role in the development of private viewdata systems.

Richard Clark was in the original British Post Office team that developed Prestel, with responsibility for its international standardization. He is now a principal consultant with Joan de Smith and Partners, although he still maintains his involvement in standards issues, representing the UK at a relevant committee of the International Standards Organization.

Keith Clarke studied engineering at the University of Bradford, and computing science at Imperial College London. After working for the UK Treasury he joined the Post Office Research Department, where he assisted Sam Fedida in drafting his original viewdata specification. He is at present Head of the Viewdata Division. He is a member of the BCS, a past vice chairman of the IEE Professional Group on Automation Applications, and is a member of the Information Processing Group.

Mervyn Grubb has had a long career in industrial distribution and latterly was Chairman of GKN Distributors Ltd. In his capacity as Chairman of the Stern Osmat Group and adviser to GKN on distribution policy, he is intimately concerned with the presentation of statistics—marketing and purchasing—in which functions he has effectively employed viewdata. An authority on inventory management, he believes the use of viewdata in this field will revolutionize stockturn in the UK.

Jaakko Hannuksela, Vice-President, Planning, of Sanoma Publishing Company, Helsinki, Finland, has been actively involved in the Finnish videotex scene since conception of the idea of Telset Project in 1976. He is a board member of the Telset network company and the Finnish representative in IVIPA, the international videotex information providers' association.

Robbie Hill, Viewdata Design Consultant CAP-CPP, has been in computing since 1963 first as compiler expert then communications. Involved in viewdata projects since 1978.

Richard Hooper was a BBC television and radio producer for nine years before entering the world of computing. From 1978 to early 1980 he was Managing Director of Mills & Allen Communications Ltd, one of the largest information providers on Prestel. On 1 April 1980 he became Director of Prestel within British Telecommunications.

Alan R. Jones is Editor, Advertising and Special Projects, at Fintel, the viewdata publishing company jointly owned by the *Financial Times* and Extel. He has been actively concerned with the international marketing of Fintel's services, both as an information provider and as an 'umbrella' services house. He has lectured widely on the advertising and promotional aspects of viewdata. He was formerly an information specialist with the *Financial Times* and with the public library service.

Folgert de Jong is a journalist working with the Dutch news agency ANP which is one of the IPs for Viditel. He had his first experience with viewdata in 1978 when he prepared and attended a joint viewdata demonstration from the Dutch newspapers and ANP.

Hisao Komatsubara has covered developments in telecommunications and related new media for the Japan Newspaper Publishers and Editors Association. As a member of the General Planning Committee of the Association of Captain Experiment Information Providers, he is deeply involved in Captain's development.

Martin Lane was involved with Prestel from its pilot trial phase when, as Business Development Manager at the *Financial Times,* he persuaded the company to sign up as one of Prestel's original information providers, through to its public service phase when, as editor at Fintel, he was in charge of one of the largest databases on Prestel. In October 1979 he was appointed Manager, Videotex Services, at Informat, the largest information service provider participating in Telidon, the Canadian videotex system.

Pat Montague was Assistant Production Manager at the *Manchester Evening News* for six years from 1965. In 1971 he joined Lancashire Colour Printers as Assistant to the Managing Director. One year later he was appointed Director and General Manager of The Birmingham Post & Mail Ltd., a post he held for four years. Mr Montague became Technical Development Director in 1976, a post he still holds. In addition, he has been the Director in charge of the Birmingham Post & Mail's videotex service—Viewtel 202—since the end of 1977.

Keith Niblett is General Manager and Director of Eastern Counties Newspapers Ltd., a group company specialising in viewdata publishing. He has been involved with viewdata since 1978, and previously was marketing manager of the company.

Hervé Nora was born in 1944 and is a graduate of the Ecole Polytechnique (1964 promotion) and the Ecole Nationale Superieure des Telecommunications (1969 promotion). Since 1 January 1980, he has been Chief of the Telematique Service, Direction Generale des Telecommunications (Secretariat d'Etat aux PTT—France).

Michael Nyhan joined the Institute for the Future after serving five years with the Aspen Institute's Program on Communications and Society. A communications and public-policy analyst, he holds a master's degree in communications and public administration (public policy development). He is co-editor of *The Aspen Handbook on the Media* and *The Future of Public Broadcasting*.

Paul Radcliffe, Viewdata Services Manager CAP-CPP, has a background of computer applications in publishing and in industrial research and development and is now responsible for a special new technology and viewdata group.

Dietrich Ratzke studied law, became a newspaper editor, editor-in-chief of a periodical, then managing editor of the *Frankfurter Allgemeine Zeitung*. In this position he came into close contact with the development of media. He wrote two books about the new media: *Netzwerk der Macht—Die neuen Medien* ('Network of power—the new media') and *Die Bildschirmzeitung* ('Newspaper on the screen'). He is a member of several international study groups. As an adviser he promoted the publishing of the first German and first Austrian 'Bildschirmzeitung'. He is responsible for the Bildschirmtext—staff of the *Frankfurter Allgemeine Zeitung* during the trial in Düsseldorf and Berlin.

Alex Reid is Director of Business Systems at British Post Office Telecommunications. From 1977 to March 1980 he was Director of Prestel. He joined the Post Office in 1972 (as Head of Long Range Studies) from University College London, where he was Director of the Communications Studies Group.

St John Sandringham has been with Consumers' Association for ten years and has been responsible for developing the 'Tele Which?' service on viewdata for the last four. With a background in consumer research and journalism, he has been closely involved in all aspects of the Prestel development and is a regular contributor on the subject.

Tim Sedman, Products Viewdata Manager CAP-CPP, has a real-time mini-computer background, and is responsible for development of telesoftware (available on Prestel) and the advanced dynamic viewdata system for training.

Colin Tipping, presently Viewdata Project Manager at Granada TV Rental, was until the end of 1978 Head of Prestel Operations at the Post Office. During his five years on viewdata he has been variously involved in such diverse elements as Prestel computer installation (for the PO) and the development of viewdata rental activities (for Granada).

Ederyn Williams took his MA at Cambridge University and D. Phil at Oxford, both in psychology. Then followed a period at University College London investigating social psychological aspects of teleconferencing. He joined Post Office Prestel in 1977, taking up responsibility for indexing the database, and for market research.

Rex Winsbury is Viewdata Director of Fintel, the viewdata publishing subsidiary of the *Financial Times* and Extel. He was formerly Features Editor of the *Financial Times,* and has worked in television and business magazines. He is author of several studies of the impact of the computer on publishing, including *New Technology and the Press,* a research report for the Royal Commission on the Press (1975); *New Technology and the Journalist,* published by the Thomson Foundation; and *The Electronic Bookstall,* a study of Prestel, published in 1979 by the International Institute of Communications. He is also author of two non-fiction books for children.

Part One

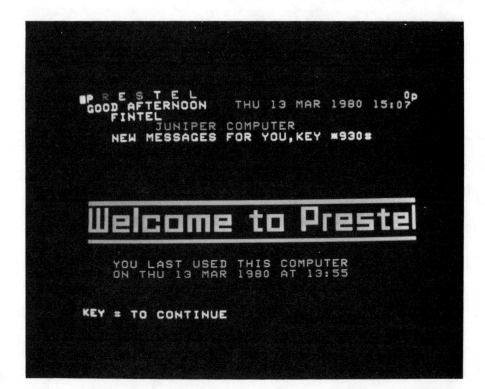

```
  P R E S T E L                                    Op
  GOOD AFTERNOON    THU 13 MAR 1980 15:07
  FINTEL
       JUNIPER COMPUTER
  NEW MESSAGES FOR YOU,KEY *930#



  Welcome to Prestel

  YOU LAST USED THIS COMPUTER
  ON THU 13 MAR 1980 AT 13:55

KEY # TO CONTINUE
```

Editor's general introduction
Six themes for study

What is known in Great Britain as viewdata, and almost everywhere else as videotex, is a system for displaying information—words, figures, diagrams—on the ordinary TV screen: the information is drawn from a computer and arrives at the TV set over the ordinary telephone line, in response to numbers pressed by the user on a keypad. That at any rate is its normal description, and will serve for the moment.

The purpose of this book is to set down the technical, financial, editorial, legal-ethical, and marketing aspects of viewdata/videotex systems at this early stage of development, drawing in particular upon the experience of Prestel, the British viewdata system that is, at this moment, several years further advanced than any other, in part by virtue of the fact that viewdata first evolved in the UK. After several years of trials, it was publicly launched in March 1980, and is already old enough to have had its drop-outs as well as its successes. Several years from now, there may be no excuse, other than patriotism, for a book about viewdata to centre upon Prestel: if it succeeds, viewdata will by then be a world phenomenon, to be considered supra-nationally. But now, more can be learnt, by all countries interested in viewdata, by a close examination of the UK experience so far, than by any other method, although this book does also contain a substantial and illuminating international section (Part Five) in which developments in most of the major viewdata countries are explained, analysed, and compared with Great Britain to bring out the differences of approach so far visible. A note on the often confusing international terminology is also included at the end of this Introduction.

This then is a book largely but not wholly about Prestel, largely but not wholly written by people involved in Prestel, but also written with a steady eye on a larger purpose, that of using Prestel as a case study, the best one so far available, of what happens when you decide to set up a viewdata system—what problems you encounter, what questions you face, what skills you have to learn, what costs you have to incur, what failures you may anticipate, where lie the best chances of success, what effect it may have on existing businesses (e.g. publishing, postal services), what technical and political choices you face, and so on. The international section (Part Five), while pointing to the actual and potential variety of viewdata in the world of tomorrow, also points to the fundamental similarity and universality that viewdata has. Lessons drawn from one case (here, Prestel) do have a universal validity, modified to a degree by local circumstances.

This is a time of intense curiosity about viewdata. Conferences about it are crowded out: there is an endless stream of visitors to and from any viewdata

3

participant or intended participant: articles, studies, consultants' reports, governmental reports, EEC reports, are beginning to fill up the shelves (although, as far as I know, this book is the first full-length attempt to set down the detail of a particular viewdata system and the experience in operating it). Viewdata projects are under way in a large number of countries, although in many of them they are at a very early stage. There are the countries dealt with in the chapters of the international section—West Germany, France, the Netherlands, Finland, Canada, the USA, and Japan—plus others not so far warranting a chapter, like Switzerland, Belgium, Spain, Sweden, Hong Kong, New Zealand. More names seem likely to be added to the list. Of the countries named, half or more have either bought the British viewdata system, or constructed one that is similar and compatible: this is a further reason why, at this stage, a study of Prestel is of wide significance.

Viewdata has, by accident or design, brought together a powerful coalition of forces in the attempt to turn it from a laboratory dream into a market reality. It is the very power of the partners in this coalition that makes it hard to believe that viewdata will not, sooner or later, carve out some useful and profitable niche for itself (although there are sceptics in plenty, and I will deal with their doubts later). This coalition is between, first, the post and telecommunications authority in each country, normally referred to as the PTT but in Britain as the British Post Office (BPO), often a national monopoly and usually on the telecommunications side a wealthy and massive national institution with an investment programme (in telephone exchanges, telecommunications networks, cables, telephones, data networks, and the rest) that is awesome in scale to humbler mortals, and dwarfs the tens of millions of pounds sterling (or national equivalent) that a viewdata system may cost to set up over 5 years or so. Secondly, there is the television industry, grown large and sometimes fat upon the boom in domestic TV and accustomed, as is the PTT, to thinking in mass-communications terms and supplying mass-produced equipment to millions of households. Thirdly, there is the publishing industry, which senses both the threat and the promise of viewdata to its traditional domain, and witnesses the arrival of a new breed of 'electronic publishers' entering the already crowded scene in what is now fashionably referred to as 'the information industry' and taking advantage of the technological developments, based on electronics, that are both transforming and widening the whole notion of publishing.

This is not necessarily a natural or inevitable coalition. The partners have not worked together before, at least not on this scale: their attitudes, economics, time scales for success or failure, resources, competitive positions, and (not least) personalities, are substantially different. This gives rise to stresses and strains, to misunderstandings, to frustrations and resentments. Yet viewdata demands that it be a coalition; demands that the three parties stand or fall together in this new market; demands that they move at least roughly in step for the venture to succeed. It has yet to be proved conclusively that they can do so: but neither has it been proved that they cannot. In the end, the prospect of making money concentrates the mind powerfully, and once past the credibility hurdle (and not everyone is past it) the tripartite nature of the venture should not in itself be a fundamental problem.

The problems are more likely to arise in the 'gutter' between the parties to the coalition—for example, in the PTT learning the sometimes vague and sometimes confused legal, social, and editorial rules and practices governing the publication of information; in the TV industry learning to think of their product as a computer terminal for text display; in the publishing and near-publishing organizations learning the economics of high-technology distribution systems. Most important of all is the question of marketing. Do the partners in the coalition market their interdependent wares separately? Or together? Who sets the tone, the timing, the scale? If it is the PTT, as the richest partner, where does that leave the others?

Here then we have the first of the major themes of any study of viewdata development—the fault lines between the major interest groupings that go to make up a viewdata system as most commonly organized today (other forms of organization are of course possible, and can be identified). There are perhaps six other major themes running through this book, and the main purpose of this Introduction is to alert the reader to them, so that he or she will pick up the echoes of them wherever they occur. It must be admitted that not all of them are treated with equal prominence or skill or depth by the contributors. This is partly because some organizations were reluctant to commit themselves to print at this formative stage, even though they are active and knowledgeable on viewdata; partly because there was no suitable person to write about certain topics—the viewdata community is still a limited one, in personality terms; but mainly because on certain key topics the evidence is as yet incomplete and inconclusive to a degree which prohibits a coherent statement of any great length. The justification for this book is that there is plenty to say, and great interest in viewdata: the penalty of catering for this interest now is that there are still some gaps. But all the main themes do find some expression—and most of them find elaborate expression.

What then are these other themes? The second (taking the coalition as the first) is the philosophy of 'the common carrier'. This is by origin an American notion, not so well expressed in British or European law, and in essence it means that the carrier of the message (in this case the post and telecommunications authority) is not responsible for its content; by extension, it also means that the carrier must carry whatever messages are offered to it for carriage, without discrimination. Thus in the USA and under the laws of the USA, American Telephone and Telegraph, more familiarly known as A.T.&T. or just 'Ma Bell', is a common carrier. What it carries *may* be against the law, for example, if it is obscene, but that is a matter for the law, not for A.T.&T., and A.T.&T. is not liable for prosecution if it carries such obscene matter, even though it may well try to prevent such carriage taking place. This notion does not exist under British law, which sometimes makes it difficult for some Americans to understand what the argument is about in Britain. In Britain the Post Office can in principle be sued in court for the matter that it carries in the mails or over the telephone system: the

5

Act of Parliament governing the BPO says nothing about 'common carrier'. But it is nevertheless a simple, compelling notion that has now become almost an article of faith (though not a necessary one) in running viewdata. The BPO has adapted it to its policy of running Prestel, and so have the Germans, Dutch, French, and many others.

This was not the original notion of the BPO: at first it wanted to manage not only the technological infrastructure of viewdata (the computers and the telephone distribution network) but also the information that resided upon the computer as well. But following objections from publishers and others, based on *publishing* conceptions of a 'free press' and 'freedom of expression and opinion', it fell back on the common-carrier concept, which rapidly became the most alluring, if at times baffling notion at the heart of viewdata development.

What it means, in a nutshell, is that the system operator, the PTT, has no control over what information is put on its computer and made available to the public, except where it breaks the law of the land. Otherwise, it is open house. Anybody, big or small, who can afford the money to rent space on the viewdata computer (or computers) is free to do so, and can put out his message to whatever audience he chooses. By this one stroke, viewdata becomes a publishing medium, and a publishing medium unlike any other, unlike conventional computer systems but, being electronic, unlike printed publishing as well. Quite quickly, viewdata becomes enmeshed with notions of a free society, of the right of anyone to speak what he will to anyone, of an 'information society' where the trade in information is perhaps the dominant activity, where the right to know is buttressed by technology as well as law. This political–philosophical dimension becomes even more striking when efforts are made to sell the Prestel system abroad to certain governments whose last wish in the world is to have uncensored free flows of information. Of course, you can control and censor a viewdata system, just like any other information source. Viewdata systems can be and are dedicated to specific purposes (i.e., display of share and stock prices) which preclude other matter; are dedicated to particular markets, like business, which preclude other matter; or are dedicated to particular publishing purposes that preclude other possible purposes. Nevertheless, for public-service viewdata systems, the 'common carrier' concept is both politically and commercially the most important and innovatory thing that they have to offer, and paves the way for almost everything else.

Thirdly, and arising from the last, there is the theme of the impact on the printed word, and in particular on the press. Is viewdata a threat to newspapers? An opportunity? Irrelevant? Will people get their news, both verbal and written, off the TV screen, along with the weather, the sports results, the TV programme schedules, and (what hurts most) their classified or 'small' advertisements as well? Classified ads are the financial bedrock of the majority of newspapers, and any inroads on that income would go straight to the jugular. The paper-making industry, in particular the newsprint industry and those companies that supply materials for paper making, like inks, coatings, etc., are equally interested in these questions, which extend also to the book, directory and catalogue publishers, and to the printing industry itself. Between them these are substantial industries and substantial employers. To find the electronics industry, computer makers, telecommunications operators, and TV industry muscling in on their territory (not, in some cases, for the first time) is an uncomfortable experience, producing some wildly silly statements both in the optimistic and the pessimistic directions.

The varied reaction of the press in particular will emerge at various points during this book. My own view is that viewdata is fundamentally a new and different medium, which either has or will evolve its own distinctive uses, some of which will in fact generate more use of paper; some of which will have some marginal displacement effect on the press (for example, in my opinion, in the case of classified advertisements); a few of which may have a massive displacement effect, such as the French scheme for replacing printed telephone directories with electronic ones using viewdata terminals; but many of which will be particular to viewdata, for example, the ability to order goods. So it would be rash *either* to ignore viewdata *or* to get hysterical about it, *or* to suppose that anyone really yet knows where, when, and instead of what it will eventually settle down as a tool of everyday life. At its simplest, viewdata is simply another opportunity for diversification for companies which are either in the publishing business or are in businesses, like computer services, which enable them to publish via viewdata.

Fourthly, there is the editorial theme—what material is suitable for viewdata, and how do you present it to best advantage on the viewdata screen? Should it be business statistics, jokes, products for sale, timetables, news, advertising, or all of those and more? Will people 'read' the TV screen with its colours and glare, and how will they know what is there to read? Viewdata is a narrow window on a huge world, with a screen display format of 24 lines by 40 characters, and not all of that can actually be used for showing information. Yet there are, allegedly, many tens of thousands of such pages of information on each Prestel computer, although as we shall see these grand totals that seem to impress the outsider are suspect for various reasons. Nevertheless, there is a real problem in: (a) breaking the information down into small lumps of this screen size; (b) stringing together these little lumps in a coherent way; and (c) letting your user know (since he can only see one at a time) what they all are.

Are we dealing with a journalist's medium where fast-changing or un-predictable material is adapted by human beings sitting at keyboards into readable viewdata pages, or are we dealing with a computer medium that uses bulk update facilities for automatic reformatting of already computerized data into the viewdata mode? Is there a synergy or a conflict between these two possible approaches? And are we to see viewdata as a medium suitable for rapid input and display of a small number of fast-changing pages, so taking advantage of the instantaneous communication ability of viewdata; or are we to see it as a medium for the storage of vast numbers of pages that do not change very often? Perhaps viewdata can do both, or more likely it will not be either a technical question or an editorial question, but a financial and commercial one as to which of those extremes or which intermediary position, proves to be most acceptable.

What skills and what background do viewdata editors need, assuming that you need any at all, and how do you train them? Are we heading for another people-intensive industry, like most of the rest of publishing, or for a capital-intensive industry where machines do most of the work? Is colour an asset or an accident, are graphics profitable or merely pretty, why do we have to talk about viewdata 'pages' at all, since the notion of a page is taken from printed publishing and is intrinsically foreign to computer systems, if not to viewdata systems? Can a computer-based system have 'style' and, if it can, what is it? These and many other editorial questions can be posed but so far only partially answered.

Fifthly, there is the commercial theme, about costs, revenues, markets, share of the cake. This is still the most elusive theme, although there is much to say about it. Viewdata may be seen as a tool for business use; a source of amusement in the home; as electronic mail order; as mass market or specialist. Each conception carries with a quite different assumption about the size of the potential or necessary audience, and about the likely spending habits of that audience. This set of variables has to be equated with another set—your costs of being on viewdata in the first place. Here you have to add together system costs (such as computer space rental, line charges, your own computer and editing equipment) and people costs, the editors, programmers, sales and managerial staff that are needed in varying combinations, along with their overheads in terms of office space, social security, effective working hours, and all the rest. So far, experience suggests that for a system that is meant to be cheap and cheerful (to use an English expression commonly applied to it), viewdata is surprisingly expensive at both ends—at the input end, to the electronic publisher, and at the output end, to the user.

But the economics of viewdata do have one attractive peculiarity for those providing the information. The cost of mounting that information on viewdata, whatever it is and however it is done, is a one-off cost that does not go up at all with the number of users or customers: put another way round, the distribution costs are borne by the user/customer, and not by the producer. This is akin to, but more extreme than, the economics of newspaper publishing, where there are very high 'first copy' costs, that is, costs to produce a newspaper at all, but costs do then rise to some degree in proportion to sales because of the newsprint, ink, and distribution expenses.

But none of that applies to viewdata. Apart perhaps from marketing and support literature, there is a one-off fixed cost for being there at all, whether you have one customer or millions. This is a formula for either losing a lot of money (at zero usage) or making a lot of money (as the graph line for revenues rises up to and beyond the flat line for costs). For those that find the magic ingredient X, there will be the sweet taste of success, and it is that which makes so many mouths water at the mere idea of viewdata.

For the other partners in the coalition, there are other commercial attractions. For the PTT, there is the prospect of generating additional telephone revenues, of business diversification, of maintaining the initiative in communications in face of the challenge from computer networks, satellite operators, reprographic companies, and a host of other communication alternatives that might reduce the traditional telecommunications business to a mere provider of lines between A and B. For the TV industry, it is a chance to expand their product range, to sell to the public the world over a new generation of TV sets to replace their colour sets, thus opening up a vista of massive new production lines and market shares changing in favour of those who grasp this multi-million-pound opportunity. Overall, there is the creation of what it is, in effect, a new industry, the viewdata industry, with its own employment and revenues, international in scope and open to all to enter at this formative period. Entrepreneurial, a gamble, unpredictable, a potential gold mine, a fine way to lose your shirt—viewdata is all of these.

Last but not least, there is the theme of technology and its likely evolution, at the same time the most important and the least important of the themes. Viewdata is a high-technology medium (its alleged simplicity is only comparative); it is technology-driven and has rightly been called a technology in search of a use. The future evolution of viewdata technology, in the direction of better screen display, especially for graphics and pictures, in the direction of local intelligence in the TV terminal, in the direction of so-called telesoftware, in the direction of international data flows—all these and more open up exciting visions of the future. At the same time it is an information medium, only one among many: the technology is secondary to, almost accidental to, the correctness, timeliness, appeal, and price of the information as perceived and used by the customer. Technology is the body, but information is the mind, even if the mind needs a body to house it.

Market demand will be the final determinant of the speed and direction of viewdata development. But the concept of viewdata is at the same time attracting many of the best minds from previously separate disciplines, in computing, publishing, information handling, telecommunications. Good ideas will be thrust into the market place to see if they prosper: indeed, it is an extension of the 'common carrier' philosophy already discussed that anyone can come along and propose his or her bolt-on goody and, within certain broad limits, have it accepted as part of the viewdata scene. Viewdata is an industry that as yet has no fixed boundaries.

Technology is the beginning of viewdata: its essential underpinning: negatively, if the technology does not work. the result is awful. Evolution of the technology will open up new possibilities, new market opportunities. But it is not technology that will, in the end, determine viewdata's success or failure, but the price and demand for the services that it has to offer.

Success or failure? No one in their right mind can be totally sure at this early stage. But for every enthusiast there is probably a doubter, a sceptic who simply does not believe that it will work. The sceptic's view, which I do not share but will try to set down as completely as I can, is as follows. Viewdata is, as we have said, a technology seeking a use. Therefore, the sceptic will go on, there is no substantial evidence of a market for it, and that is a bad way to do business. What is more, viewdata will for a long time be too expensive for the average household, thus denying the mass market upon which PTTs and TV industry are putting their money. But for business, where cost as such is not so important provided the product is right, there are plenty of competing systems of information storage and transfer, compared to which viewdata is a crude and primitive instrument. Therefore the smaller but theoretically lucrative business market will be denied as well.

As long as newspapers and magazines are so cheap, thanks in large part to the subsidy they receive from advertising, and as long as so much television is virtually free, again thanks to the revenues from advertising, viewdata will be hopelessly uncompetitive, on this view. Ordinary people are not used to paying for informatiom, and business, which is, wants something better. And what is there that viewdata can do that cable systems, colour video display units, smart copiers and smart typewriters, interconnecting word processing systems, and video games cannot do as well or better?

Then there is the competition for the consumer's spare pound and spare dollar—can the average household or even the not-so-average household be expected to purchase a video cassette, and a video disc machine, and video games, and a second or third TV set in the house, and teletext services (those cousins of viewdata explained in the note on terminology), and music centres or racks, and calculating-alarm-clock-radios, and viewdata, to name but a few? No doubt one of the avenues of technological advance will be to start marrying these various functions into one box, as cassette-radio-alarm-clocks already do and as viewdata sets that also receive teletext already do. Nevertheless, there is a limit to disposable income and the customer has to choose between alternative satisfactions. On that scale, the sceptic reckons, viewdata with its attendant expense comes out low on the priorities.

Viewdata was recently compared by one critic to someone who has invented a square ball, built a stadium round it, and is now looking for a game to play. That just about sums up the sceptic's case. But the force of the sceptic's case—for force it has—can be counterbalanced by the power of the coalition mentioned above, by the range of countries that have taken an interest in viewdata, described in the international section, and by the simple attractiveness of the medium itself. It is said with great truth that the easiest way to sell someone viewdata is to sit him (or her) down in front of it. It is hard for total scepticism to survive the simple impact of seeing well-designed viewdata pages, serious or amusing as the case may be, roll by on the screen at the simple press of a button on a keypad.

Viewdata is also part of a wider development that takes in teletext as well. Put the other way round, the movement towards using the TV screen for text display in the home and office has two parts to it. One is the transmission of that text to the screen by the ordinary broadcasting networks: the other is the transmission of that text to the screen by telephone line from a computer. The former is known, in the UK, as teletext: the latter as viewdata. Just as viewdata has a 'brand name' of Prestel, so also teletext has two brand names, Ceefax and Oracle, used respectively by the British Broadcasting Corporation and the Independent (commercial) television network. The close relationship of these developments is recognized by the use of the international term 'videotex' to describe both, broken down into 'broadcast' and 'interactive' videotex. So far, teletext has not spread widely in the UK, although it has been on the market for about 5 years: it has not been the subject of any substantial promotion by the broadcasting companies. Whether teletext will be borne along by the development of viewdata, will be a competitor to it, or will ride in tandem with it, is one of those so-far tantalizing questions. Teletext is not treated in this book, since in the UK at least it is for the moment the sole prerogative of the broadcasting companies to operate teletext services, and until they decide to put their weight behind teletext, there is not a lot to say about it of general interest except that it exists, is free to the user, and could do great things. But in assessing viewdata, it is important to remember the broader picture of which it is part.

What makes viewdata in particular so intriguing, and what raises so many questions about it, is that because of its specific combination of technology and participants, it sits at the cross-roads and cross-over-point between paper-based publishing and electronic publishing. The medium is new, but many of the editorial and marketing decisions are more akin to those of traditional publishing, in the sense of what information to market, to whom, and at what price. Similarly, the viewdata page, with its mixture of colours, graphics, text, figures, indexes and routing, is a unique amalgamation of elements drawn from previously separate disciplines, making it a new creation, a new vehicle of expression. The attraction of viewdata lies in the challenge it presents to make creative use, both editorially and commercially, of its unique properties.

Terminology

Unfortunately, there is no internationally accepted terminology to describe what is going on. The British system and its participants have always used the terms 'viewdata' and 'teletext'; but more recently the International Telecommunications Union, which is the forum for the standardization of these matters, has provisionally adopted the word 'videotex'. The terminology can be explained as follows:

Videotex is the generic name used in many countries for electronic systems that use a modified TV set to display computer-based information. Interactive systems using typically, the TV set and telephone line, are called telephone-based or interactive videotex. Broadcast services, in which the pages are sent over the air waves, are called broadcast videotex.

Viewdata is the UK term for interactive videotex, and is the term used in this book.

Teletext is the UK generic name for broadcast videotex services, and is the term used in this book.

Prestel is the proprietary 'brand name' of the interactive telephone-based videotex (viewdata) service of the British Post Office. Many other countries also have such brand names, e.g., Viditel in Holland, Teletel in France.

Ceefax is the broadcast videotex (teletext) service run by British Broadcasting Corporation.

Oracle is the broadcast videotext (teletext) service run by the independent television companies in the UK.

Information Provider (IP for short) is the accepted term for a publisher on Prestel.

PTT means the national postal, telegraph and telecommunications authority.

Viewdata sits between two rival jargons, two competing private languages. One, the computer language, speaks of a database maintained by manual or automatic update, with operators keying in, or bulk input by magnetic tape transfer or computer-to-computer handshake, with data reformatted to fit the Prestel display mode and generate the indexes.

The other, the publishing language, speaks of a magazine or information service published on viewdata, with news, advice and useful figures prepared and kept up-to-date by editors and presented on Prestel pages with suitable layout and cross-reference, using editing keyboards.

There are a host of associations and assumptions that go with each language. Neither is wholly appropriate to viewdata but both are partially appropriate. That is one measure of viewdata's novelty and challenge. Both languages are used here, inevitably if sometimes confusingly.

1. Prestel philosophy and practice

Alex Reid

Invention is the right place to start. For the Prestel service, and the viewdata industry of which it is a part, sprang from an idea: Sam Fedida's idea that TV sets should be linked through the telephone network to computerized stores of information.

With Keith Clarke and the other members of his team at the Post Office Research Centre, Sam Fedida developed this concept during the mid-1970s to the point of a practical working system. During the same period, engineers at the British Broadcasting Corporation and the Independent Broadcasting Authority had devised two related systems (respectively christened Ceefax and Oracle) which similarly linked the TV set to a computerized store of information—but in their case this was done using the one-way wireless broadcast signal, rather than the two-way telephone network. Collectively, Ceefax and Oracle are known as teletext services; firm technical standards for teletext were agreed by the BBC, IBA, and BREMA (British Radio Equipment Manufacturers Association) in 1976. These technical standards (24 rows of text with 40 characters per row, in any combination of 7 colours) produced on modern colour TV sets an excellent display, that was lively, crisp, and legible. The Post Office, which had hitherto been working on an entirely different technical standard (13 rows, 32 characters per row, with no colour or graphics), quickly recognized the merits of the teletext display format and adopted it. This commonality of display format had the advantage that TV sets designed to receive both teletext and viewdata could, to a considerable extent, use the same electronics for both services.

The Post Office Telecommunications Business

Until the spring of 1977, viewdata was a research project. In 1974 and 1975 it had been widely demonstrated, and in 1976 a pilot trial started, with a few information providers and a few dozen viewdata sets in their premises and those of the Post Office.

It was in the spring of 1977 that a proposal was prepared, by the viewdata project team, for a viewdata market trial involving up to 30 000 pages of information, and 1000 viewdata sets in homes and businesses. The viewdata sets were to be manufactured, at their own expense, by TV-set manufacturers and were to be rented out to volunteers in London, Birmingham, and Norwich. The Post Office would have to spend £3.5 million to establish and run the market trial for one year.

This proposal was approved first by the Post Office Telecommunications Business, and then, in March 1977, by the Post Office Board. This decision was influenced by four main factors. Firstly, the product itself was very appealing—

almost hypnotic when well demonstrated. Secondly, there was already tangible support for viewdata from British TV-set manufacturers (who were willing to provide, at a subsidized price, the 1000 viewdata sets for the market trial) and from information providers. The list of prospective information providers to the market trial was particularly impressive—well-known newspaper and magazine publishers, leading private firms and many government and non-profit-making organizations. Much of the credit for attracting these people must go to Malcolm Smith and Roy Bright of the Post Office Telecommunications Marketing Department, who gave innumerable demonstrations of viewdata in a circular demonstration room in the basement of Lutyens House, Finsbury Circus, London. The third factor was that the Post Office Telecommunications Business was conscious that its good record of productivity improvement and harmonious industrial relations owed much to the enormous rate of growth in telecommunications; if this growth was to be sustained over the next twenty years, new services such as viewdata would have to be created. A particular attraction of viewdata lay in the fact that it would generate telephone traffic and extra revenue through more intensive use of the existing telephone network. Moreover, after a period of heavy losses caused by anti-inflationary tariff restraint, the Post Office Telecommunications Business was now producing a good financial performance, and was in a mood to expand.

Simple, cheap, universal

The inherent nature of the product, when married with the objectives of the Post Office Telecommunications Business, produced a clear-cut market strategy. If the service was to grow to a scale that would make a significant contribution to the multi-billion-pound telecommunications business, it must eventually be a large national service comparable to the postal or telephone services. To achieve this, the service would have to be simple, cheap, and of universal application in homes, offices, shops, factories, and schools. This was exactly the mood of the product itself, for viewdata brought together the TV set and the telephone—two simple, cheap, and universally available devices.

This breadth of approach was in bold and conscious contrast to previous computer-based information services, for example, for airline-seat reservation, share prices, betting offices, bibliographic search, or currency exchange rates. Each of these services had been custom-built for a very specific and limited purpose. They are powerful and ideal for their purpose. But they are expensive and specialized; too expensive and too complicated to grow into an everyday national information service.

Although the product strategy—of simplicity, cheapness, and universal application—was clear cut, it was by no means obvious how such a strategy could be brought to profitable realization. Our analysis of the commercial problems led us to three conclusions:
- we needed allies
- we could not afford to have enemies
- we should move fast.

Allies

We needed allies because, although Post Office Telecommunications is a large and powerful organization, it possesses only a fraction of the skills and resources needed to launch and operate a national viewdata service. Given the central role of the TV set in viewdata, it was essential to recruit as allies the TV-set manufacturers and distributors—including the TV rental companies, which play such a large part in British TV-set retailings. And because the essence of viewdata is the *information* it offers, we had to recruit as allies organizations which owned information of all kinds, and knew how to present it effectively.

Enemies

The absence of enemies was just as important as the existence of allies. In several foreign countries, the development of viewdata has become bogged down in political and regulatory wrangles, and if we had made enemies the same thing could have happened here. There were three potential groups of enemies:
- firms who perceived viewdata as a potential threat to their established line of business, e.g., newspapers, publishers, commercial broadcasters, computer bureaux, and TV-set retailers.
- organizations that wished to act as information providers to viewdata, but either were not allowed to do so, or were unacceptably restricted in terms of editorial freedom or pricing.
- politicians, academics, journalists, or others who felt that viewdata was being handled in a way that was socially or politically undesirable. There would be grounds for such objections if, for example, the Post Office unfairly favoured particular providers of information, unfairly excluded others, or interfered with the traditional freedom of the press.

A commercial formula

Our commercial formula to attract allies and to avoid enemies has three elements:
- adopting a common-carrier policy towards information
- creating a free market in viewdata apparatus
- welcoming competition with ourselves.

Because these policies are so fundamental, we must look in detail at each; first describing each policy and the reasoning that led to it, then setting out its practical implications.

Why the common carrier?

Perhaps the most fundamental policy issue with which viewdata presented the Post Office was that of editorial control. Should the Post Office select the information providers, and then control their editorial activity? Such an approach would be similar to that of the Independent Broadcasting Authority, which selects and controls the commercial TV programme companies. At the other extreme, should the Post Office run Prestel as, in effect, a common carrier renting space in its computers to anyone, and leaving to the information

15

providers complete responsibility for the information content? Such concentration on the *carriage* of information, without regard to its *content* would be in the tradition of the postal and telephone services. Or should the Post Office adopt some intermediate policy, such as selecting the information providers but giving them complete editorial freedom?

In the event, we plumped firmly for the 'common carrier' approach. We rent space in our Prestel computers to anyone who can pay the storage charges. The information providers may choose whether to charge the users to look at their information, and if so, how much; and information providers may put up on their pages whatever information they wish, subject only to the law. There were three reasons for adopting this policy—political, administrative, and commercial.

The political reason was that we felt a policy based on the Post Office choosing information providers, and then exerting control over the information they provided, would embroil us in political controversy. Rejected information providers would protest at their loss of commercial opportunity, and those who were chosen would find it unacceptable to have the Post Office as a kind of grand editor and censor of what they could publish. This latter point was brought home to us by the strongly and persuasively expressed views of some of the leading prospective information providers. They pointed out, very reasonably, that the press had fought a long and successful battle for editorial independence. They saw no reason why that independence should be sacrificed just because they were printing on the TV screen instead of paper.

Beyond the feelings of individual information providers, editorial control by the Post Office would raise wider issues for political debate. What balance should be struck between commercial and public-service information? How many pages should be devoted to the Welsh language? What references could be made to tobacco, alcohol, and drugs? What constraints should be placed on the treatment of politics, religion, and sex? In no time there would be call for a Government Committee of Enquiry, which would involve considerable delay, and would probably lose Britain its world lead in this fast-moving and competitive technology.

Generally, we were conscious of a world-wide trend towards greater pluralism and diversity in the communications media. It takes many forms: the trend towards hundreds of specialist magazines instead of a few dominant national weeklies; the burgeoning growth of local radio; the interest in cable TV and community video; flexible TV viewing through video cassettes. Any attempt by the Post Office to exercise central control over the content of viewdata would have flown in the face of these trends. By contrast, the common-carrier policy is a spectacular demonstration of pluralism and diversity in action.

These were the political reasons for adopting a common-carrier policy. The administrative reason is pretty obvious when you consider the effort that would be required to monitor and control hundreds of thousands of pages (eventually millions of pages), each one of which can be changed many times a day by an information provider from a terminal in his own premises. The control mechanism would not only have to be large, it would have to be complicated and wide ranging. Like the Independent Broadcasting Authority (which has the relatively manageable task of controlling a single TV channel and a few dozen radio stations), we would need national committees, religious committees, regional committees. We would need experts in good taste who could pronounce on the acceptability of specific references to vicars, Irishmen, or parts of the body. We might even need our own Egon Ronay and our own Bernard Levin to ensure quality in our restaurant guides and theatre reviews.

All this was an administrative task which we were glad to be shot of. We are, however, legally liable for the information appearing on Prestel. This imposes on us a minimal but inescapable regulatory role. Briefly, our information providers undertake in their contracts with us to indemnify us for any damages awarded against the Post Office on account of the information providers' pages. If we judge that an information provider's information is committing a criminal offence, we reserve the right to bar access to such pages until the matter has been legally resolved. The administrative effort needed to operate these legal safeguards seems to be mercifully small, perhaps because 99 out of 100 people are anxious to avoid breaking the law.

In addition to the political and administrative arguments, there were commercial reasons for adopting a common-carrier policy. The Post Office does not claim to have commercial experience of, or commercial flair for, the publishing business. If we had shaped the content of viewdata, the service would inevitably have reflected our own interests and tastes, which would not necessarily coincide with the right formula for commercial success. By contrast, the common-carrier policy gives full rein to the commercial interests, motives and flair of the established publishing industry. Because a common-carrier policy gives no exclusive rights to any information provider, it encourages competition in quality and price. We can already see in Prestel fierce head-to-head competition between providers of similar information such as sports results, railway timetables, and property and job advertisements.

Consequences of the common-carrier policy
The common-carrier policy has several practical consequences. One of them—the need for legal safeguards—has already been mentioned. Four others are the problems of page numbering, of indexing, of incentives, and of database growth.

Each page in Prestel carries an identifying number. The choice of these numbers is a non-trivial task, because subordinate numbers (such as 65 644) route naturally from shorter numbers, higher in the hierarchy (such as 6564 and 656). Our first approach was to assign the short, high-level numbers to particular topics, such as gardening or shares, so that all the pages on that subject from all information providers started with the same three digits, and sat within a rigid numbering scheme. We could see that this approach would lead us into difficulties, and in 1977 we changed tack and assigned a three-digit number to each information provider, with all his pages (on whatever subject) grouped under that number. This has produced two enormous advantages. Firstly, it enables the information provider to rent a block of indifferentiated pages and gives him complete freedom to change his subject matter at will. Secondly, it gives each information provider a snappy and memorable three-digit front door to his database. These three-digit numbers are being vigorously promoted in the same way that a commercial radio station promotes its wavelength.

One penalty of the common-carrier policy is that it throws up a massive and complicated indexing task for the Post Office. The information providers have produced material of enormous variety—share prices, Bible readings, encyclopaedias, horoscopes, fruit machines, muck-spreading prices, population statistics; even (my favourite) an advertisement for an electronic ferret locater at the reasonable price of £24.50. The Post Office indexing team have the formidable task of leading the users quickly and easily to the needle of information they need in this haystack of 150 000 pages.

We examined past practice in book indexing, library cataloguing, and searching of bibliographic databases, but found that we really needed a new approach. The methods of book indexing are inappropriate because our technology is different. The methods of library cataloguing and bibliographic database searching are inappropriate because both tend to rely on skilled staff to do the searching, whereas Prestel is intended to be used by people without any such training.

We have therefore had to devise new methods of indexing for Prestel, appropriate for the technology and for the user. To help us, we commissioned research studies at Loughborough University and the Post Office Research Centre and got advice from ASLIB (the Association of Special Libraries and Information Bureaux).

There are two types of index—on the screen and on paper. The screen indexes consist of about 1000 pages offering the user the choice of searching by simple dialogue (leading from very general choices to successively greater detail), by an alphabetical index of information providers, or by an alphabetical list of topics. The paper indexes comprise alphabetical listings of information providers and of topics. They are compiled by the Post Office, and are incorporated in the three commercially produced Prestel directories, which are provided free to Prestel users. The three basic issues in constructing both sets of indexes are: How complex? How deep? How evaluative? On each question, our approach is pragmatic. I would describe our indexes at this stage as fairly complex, fairly deep, and fairly evaluative. On each aspect, we are learning as we go.

The common-carrier policy looks to the discipline of the market to encourage the good and drive out the bad. Market discipline requires effective feedback from consumer to supplier, and we are doing all we can to amplify that feedback. The computer tells the information providers how often each of their pages has been looked at, and we publish on Prestel a 'Top Ten' league table, showing which information providers get most accesses. We have a substantial market-research programme (including about 1000 test service users) which provides more detailed audience reaction through personal interviews. Prestel users can send compliments and complaints to information providers electronically, using a Prestel response page, and we publish (on Prestel) league tables showing which information providers got most of each.

A final consequence of the common-carrier policy is the necessity to increase the capacity of the database in order to meet all the demand from information providers. Originally designed in 1977 to hold 30 000 pages of information (which seemed a lot at the time), the capacity of the database has been progressively increased to 60 000 pages, then to 120 000, and now 250 000. A decision was taken in September 1979 to double again to 500 000 pages, but this will not be technically practical before late 1981. There has been tremendous demand from information providers for database capacity and we are at present completely sold out. This has forced us into the position of allocating such pages as they do become available to those information providers who seem likely to make the greatest contribution to the commercial success of Prestel. This selection is regrettable, but commercially necessary during a period of database scarcity. Our aim is to expand capacity to get ahead of demand, so that we can revert to the common-carrier policy of accepting all comers.

Why competition in terminals?
The second plank in our communal formula is a liberal approach towards the attachment of Prestel sets to the telephone network. Subject only to technical tests of each type of set (to avoid damage to the telephone network), any organization can make and market Prestel sets. Such sets can contain all the electronics needed to connect Prestel sets to the telephone line—memory, character generator, decoder, terminal identifier, autodialler, and modem.

Hitherto, the Post Office had exercised its statutory monopoly over the supply of modems (which connect data terminals to telephone lines) and autodiallers (which automatically call the computers telephone number). In respect of Prestel, we have relaxed that monopoly. The manufacturers are also free to add special features of their choice, such as hard-copy printer, memory intelligence, broadcast reception, alphabetic keyboard, audio cassette recorder, or coin mechanisms for sets in public places.

The reason for encouraging private firms to make and market Prestel sets (rather than adopting the monopolistic approach of supplying all Prestel sets from the Post Office) is that Prestel uses the new TV set, and TV sets are already supplied by private firms. Our policy does not exclude the Post Office from supplying Prestel sets (in fair competition with the private sector) and a market trial of Post Office-supplied monochrome business sets is being held in Scotland.

The reason for allowing the TV-set industry to incorporate modem and autodialler within their sets is that the alternative would have been the commercial death of Prestel. For if every Prestel customer was forced to have a separate modem, and a separate autodialler, the result would be clumsy, untidy, and much more expensive.

There are two reasons for allowing the Prestel set makers freedom to add extra features to their sets. The negative reason is that the TV-set industry is complicated and competitive, with new TV features (such as remote control and teletext) appearing all the time; the manufacturers would have been reluctant to include Prestel facilities in their sets if this restricted their freedom to add such features. The positive reason is that Prestel itself is a ferment of innovation. It is by no means clear just what variations of the Prestel set are wanted by the market; our liberal policy encourages the manufacturers to experiment and to compete with each other in features as well as price.

Competition with Prestel

The third plank of our commercial formula is that we allow, indeed we encourage, the establishment of other viewdata services. The Prestel set can store several telephone numbers, and can therefore call other viewdata computers as well as the Post Office's Prestel computers. These private computers could offer viewdata services which were complementary to, or in direct head-to-head competition with, the Prestel service.

Our reasons for this open approach are both commercial and political. Commercially, our key task is to persuade people to buy or rent Prestel sets; the more services that are available, the more useful and attractive these sets will be. Politically, we foresaw strong and understandable objections to a restrictive policy. For an attempt on our part to arrange that Prestel sets could only call our Prestel computers would be objectionable both to users (since it would reduce the usefulness of their sets) and to operators of private computers (who would be shut out of the Prestel market).

Consequences of competition with Prestel

The Prestel service is, unlike most Post Office telecommunications services, in the competitive arena. This has some practical consequences, of which we focus here on three: pricing policy, investment planning, and trade mark protection.

Our pricing policy has as its starting point that Prestel is a separately accounted profit centre within Post Office Telecommunications, and must set its prices at a level which will cover its identifiable costs and will earn an adequate return on the capital it employs. No account is taken, in these pricing calculations, of the revenue which the main Post Office Telecommunications Business gets from the telephone calls to Prestel. Such revenue is excluded from the Prestel accounts because it is needed to expand the telephone system, and because such cross-subsidy would enable Prestel to compete unfairly with private viewdata services.

The next principle of our pricing policy is that Prestel prices should clearly distinguish and reflect the separate elements of cost, i.e., the computer capacity needed to handle users, the computer capacity needed to store data, the provision of information, and the costs of billing. The first of these elements is reflected in the time-based usage charge (currently 3p for one minute in the business day, otherwise 3p for three minutes). The second is reflected in the storage charges paid by information providers (currently £4000 per annum plus £4 per page per annum). The third element is reflected in the information providers' page charges (from 0p to 50p per page, shown at the top right-hand corner of each page). The fourth element is reflected in the 5 per cent of information providers' charges which the Post Office retains as a contribution to billing costs.

The final twist to our pricing policy is that in order to encourage off-peak use of service, we have the reduced usage charge outside the business day, and we waive, for residential customers, the £12 per quarter standing charge. In summary, our competitive situation forces us to adopt a keen pricing policy, closely reflecting costs.

Competition also has important consequences for our investment policy. At one extreme, we could have sought to bring the service rapidly within local-call reach of the whole population, regardless of cost. At the other extreme, we could have concentrated the service initially in London, sat tight there for some years until we had shown a profit, and only then extended the service to other cities. Competition rules out the first policy, for it would be impossible for us to compete successfully in the profitable urban areas, with private firms which concentrated on those areas alone. We rejected the opposite extreme, of concentrating in the early years on London alone, because we thought it was unnecessarily timid.

Our investment policy strikes a middle course between these extremes. At the time of writing (September 1980), Prestel is available on a local-call basis in London, Birmingham, Nottingham, Manchester, Liverpool, Leeds, Newcastle, Edinburgh, and Glasgow. This represents 48 per cent of UK residential phone lines and 48 per cent of UK business phone lines. By the end of 1980 we hope to open enough new centres to bring Prestel within local-call reach of more than half the phones in the UK.

Our competitive situation had important consequences, too, for our policy on the naming of the service, and its trade mark protection. The word 'viewdata' was coined in the early 1970s in the Post Office Research Centre. But it is not registrable as a trade mark, because it is made up of two English words descriptive of the product. It is not generally realized that the inventor of a thin cake cannot register as a trade mark for that product the name 'Thincake'. But the name 'Thincake' could, in principle, be registered as the trade mark for any other product, e.g., an orange or a gumboot. So viewdata (always with a small 'v') became the generic descriptive term for services of this kind. But alongside, and complementary to, that generic term we needed a registrable trade mark for our viewdata service, for the same reasons that Kodak, Coca Cola, Shell, and Rolls Royce need protectable names under which to conduct their business.

We needed a name that was:

- registrable as a trade mark in the UK and abroad (i.e., not descriptive of the product, and not already in use);
- devoid of misleading or unfortunate associations in English and foreign languages;
- short, memorable, and easily spelt and pronounced by speakers of English and other languages.

During the selection process some 200 candidate names were tested for registrability, in several countries, and were submitted to consumer research into

21

their pleasantness and meaning in several languages. Several of us in the project contributed candidate names. Others were suggested by our consultants in this exercise, Novamark Ltd, whose sole line of business is the devising and testing of brand names. Yet others were generated by computer. A list of some of the names which were considered is in Table 1.1. Optel and Prestel were two suggestions of my own. I preferred Optel, since it combined with syllable 'tel' (which has the Greek meaning 'at a distance') with the syllable 'opt'. 'Opt' has two distinct Greek meanings; 'to see' (as in optical) and 'to choose' (as in option). Both are highly relevant to Prestel which enables you to see what you choose (at a distance). This somewhat abstruse scheme fell to the ground when we discovered that Optel was already in use as a trade mark for electronic equipment in the USA. Prestel does not have the same linguistic ingenuity, but is in every practical respect an excellent name, and it emerged from the selection process as the clear winner.

Table 1.1 Examples of words which were considered as proprietary names for the Post Office viewdata service

autodata, autodial, autofact, autonews, autovision, codenews, datac, datacode, datafind, datafact, datakey, dataview, datax, dialafact, dialex, findex, findout, infodial, infoscan, infoview, instacode, instadata, instadial, instafact, instat, keydata, keycode, keyscan, newscode, newsfact, scandial, telecode, telet, teleview, videogen, videoscan, viewdial, viewex, visionex, vistac, vistacode, vistafact, vistan, vistaview, vistax, contact, scope, focus, folio, update, insight, line up, on tap, newsline, branchline, pacer, tracer, trailer, reporter, primer, primus, opus, witness, telefile, telfile, telescan, telscan, epoch, fast facts, whoosh, swift, info, express, winner, flash, mercury, zeus, midas, ulysses, sunny, finder, commander, helper, aider, fingertip, keyfax, daydata, maxidata, premier, hotline, home line, immidata, striker, contact, compress, victor, keyright, touchline, excel, servant, shortcut, lightning, superline, hermes, minder, merlin, lexiphone, anchor, searcher, wizard, factfinder, plato, producer, solomon, aspect, prospect, visor, eureka, factel, telect, beeline, optel, intouch, mentor, quest, signet, precept, signpost, redline, checker, saga, prelude, genius, interview, meractor, motive, dialogue, mandate, solve, evolve, limelight, castel, citation, ino, avenna, red pages, octave, pyramid, digit, enifact, into, optic, pointer, presto, tell, title, blueprint, checkline, citel, telo, vudata, vudatel, viewdatel.

Having chosen a name, we needed to embody it in memorable visual form. We commissioned the Pentagram design consultancy, to produce a house style for Prestel, including a 'logo' which could be widely used as a symbol, or badge, for the service. The task was a tricky one, for the logo has to look good on paper, but had also to be displayed effectively on the Prestel screen (both as a complete symbol and as a single line on the top left-hand corner of all our index pages). After some false starts, Mervyn Kurslansky produced the solution shown in Fig. 1.1.

This can be used either in the monochrome form shown here, or in colour. In colour the diamond remains black, but the letters of the word PRESTEL are respectively green, red, cyan, blue, white, magenta, and yellow (the seven colours of the Prestel screen). And the middle line of the logo can be extracted to be used as a single-line identifier at the top left-hand corner of each Prestel index page.

Fig. 1.1   Prestel Logo

The brilliance of this solution lies not only in its practical effectiveness, but also in its intellectual elegance. The black background corresponds to the black background of the Prestel screen; the seven letters of the word 'Prestel' tie in with the seven colours on the Prestel screen; and the unusually wide spacing between the letters reflects the need, in the original Prestel coding system, to leave a blank space when changing colour.

Putting the commercial formula into effect

To recapitulate the three elements of our commercial formula, they are:
- a common-carrier policy towards information
- creating a free market in Prestel apparatus
- welcoming competition with Prestel.

The formula was designed to gain us allies, and to avoid creating enemies. It seems to have been successful on both counts. On the first count, we have been joined in the Prestel venture by virtually all the leading firms in the sectors with which we are concerned—electronic components, TV-set manufacture, TV-set distribution, telecommunications, computing, and publishing.

When one looks back at the commercial forces that gathered to resist earlier media innovations (such as radio and cable TV), we have remarkably few enemies. Some organizations, such as the Advertising Standards Authority and the British Medical Association, have expressed anxiety at possible abuses of the editorial freedom which we allow to information providers. And some organizations concerned with community and social information have been worried that commercially profitable applications of Prestel would drive out the non-commercial ones. But these anxieties have been relatively mild, and the organizations involved seem to accept that while no policy can fully satisfy everyone, we are proceeding in a reasonable way.

Speed

There are tremendous economies of scale in Prestel; for the Post Office, which incurs large fixed costs in running any Prestel service; for the chip and TV-set

23

makers, who need to spread development and tooling costs over many customers; and most of all for the information providers, for whom costs are fixed, but whose revenues are proportional to the number of Prestel users. For all of us, therefore, it is vital that Prestel grow quickly to a size that will be commercially viable. There is no magic number of users; some firms will break even when there are 10 000, some need 100 000, and some need a million or more.

How to grow quickly? For us in the Post Office, there are four main things that must be and are being done:

- *Early commitment to a public Prestel service.* We committed ourselves to a public Prestel service in February 1978. Now, more than two years later, no other country except, perhaps, Finland has yet committed itself to a public viewdata service.
- *Avoiding technical grandeur.* Our system design is modular and robust, based on a growing family of mini-computer installations—not on any grand scheme for an enormous and complicated central main frame computer.
- *Acceptance of technical risk.* Our operations division is deliberately pushing technology into public service at the earliest possible date, even if this means that a lot of the debugging has to take place after, rather than before, the machines are brought into public service.
- *Vigorous marketing.* Our marketing plans include heavy use of TV and press advertising, conferences and exhibitions, activation of high-street TV showrooms, and the commissioning of intensive selling efforts in specialist markets such as investment, property, the racing fraternity, and the rich.

Speed of progress depends also on the efforts of the information providers and the TV-set industry. The information providers have shown a wide range of performance, from the highly professional and effective efforts of the leading firms to some at the other extreme who have made virtually no use of their pages. At the time of writing, the TV-set industry have not yet performed in terms of volume delivery and sale of Prestel sets, with only about 200 per month being added to the system. But a sharp upturn in supply seems imminent. Four leading manufacturers estimated that they would be producing Prestel sets at a combined rate of hundreds per week before the end of March 1980. It was for this reason that the Post Office felt able to plan a substantial promotional campaign for Prestel, including heavy TV advertising, for the last week of March 1980.

The world our oyster

Ever since Prestel entered the pilot trial stage, in 1977, we have taken a keen interest in:

- international technical standards for viewdata
- the scope for sale of our world-leading Prestel technology to foreign telecommunication administrations
- the opportunity to offer an international Prestel service from computers based in the UK.

On technical standards, we have been putting the case in the international committees (CEPT, CCITT, CCIR, etc.) that technical standards for viewdata should be broadly compatible with those already established in the UK viewdata and teletext services. Our standards are technically sound, and they provide a firm foundation on which to build future enhancements such as improved graphics and messages.

On the overseas marketing of our technology, we have already sold viewdata software and know-how to telecommunications administrations in West Germany, the Netherlands, Switzerland, Hong Kong, and (via our licensees Aregon—formerly called INSAC) the USA.

On an international Prestel service, a market trial is being aimed specifically at the international business community. Managed by our consultants, Logica, it will connect terminals in the UK, Sweden, West Germany, Switzerland, Australia, and the USA, to a specially assembled database of international appeal. It will give us the evidence of demand and feasibility on which we can base a decision whether to proceed with a full commercial service. Thus Prestel in the 1980s should be both a national and an international service.

Part Two

Technology and industry

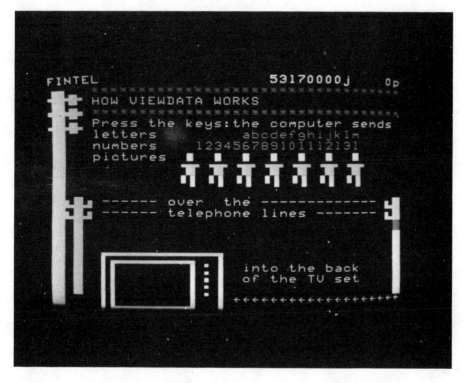

Editor's introduction
A lay guide to how it works

The following chapters deliberately aim at the expert audience, providing an authoritative account of how viewdata works in a technical and computing sense; what future developments there might be; what problems there have been in attaining international agreement; and what the reaction has been from the television industry. Perhaps this then is the appropriate place for a brief *layman's* description of how it works, for the non-experts, so that those who cannot or chose not to take in these chapters, can nevertheless read the remaining sections of this book with adequate understanding. The expert audience can skip on.

Viewdata revolves around one simple notion, that of linking the ordinary television set to the ordinary telephone line, so that the TV set can display on its screen not just television programmes, but text, figures graphics, and pictures. All these are brought to the TV screen over the telephone line from a computer. The user of the TV set has a small keypad which has on it, besides the controls for the TV programmes (channel selector, brightness and contrast controls, volume control, etc.), a set of buttons like those on a pocket calculator. These are labelled with the numbers 0 through to 9, and the user presses these to get what he or she wants from the computer. Many of these keypads are connected to the TV set, not by a wire, but remotely by audio bleep, infra-red or other signals.

When the user wants Prestel, for example, he or she normally presses the special 'viewdata' button on this keypad, and this activates an autodialler inside the set (which has inside it some extra electronics). This dials up the telephone number of the computer, indentifies to the computer the TV set and the unique pass number of the user (so that the computer knows who to send the bill to) and then the computer replies by displaying a welcome notice on the screen and then asking the user what he wants.

The user makes his or her choice in either of two ways. The first is that the computer displays on the screen an index, with numbers against the index entries, and the user presses the appropriate number on the keypad. Usually this will bring from the computer a second, more refined index that asks the user to make a further choice, and this process may repeat itself several times before the user finally gets to the information that is sought—say, the weather forecast.

The second way is that each of the pages on viewdata (they are normally referred to as pages, and for the purpose of this description a 'page' means a screenful of information, the amount displayed on the screen at any one time) has on it a page number, somewhere between 0 and 999 999 999. If the user knows the page number of the information wanted—for example, knows that the weather forecast is on page 10 101, then all those indexes can be avoided by keying that page number on the keypad, along with two special control keys known as the 'star' and the 'square' (or 'hash'), and the computer immediately sends over that page. This is the most economic way of using the viewdata.

Each page may be left on the screen for as long as the user wants, and will stay on the screen even when the telephone connection is broken, until such time as the TV set itself is switched off. This is because there is a local page store or buffer inside the TV set, which in fact stores 4 pages at a time. This means that the user can, by pressing the right combination of keys on the keypad, go back 3 pages from the one he is at present looking at.

The display on the screen is usually in the seven colours that go to make up a normal television picture, but as with TV itself there are also black-and-white Prestel sets. Usually the Prestel set is a normal TV set receiving the standard TV channels, but there are special Prestel-only sets. If the Prestel TV set is connected via your ordinary telephone, then the telephone cannot be used at the same time for other calls, since the telephone line is occupied.

Because viewdata is a system that relies on the telephone line (again, usually), it is normally the telephone company itself, like the British Post Office, which operates viewdata. But other organizations can and do set up viewdata systems. The user normally pays for viewdata in three ways: for making the telephone call (unless he or she is in a country where individual calls are not normally paid for); a special access charge to get into the computer; and for the information actually read on the screen. Each page has a price displayed in the top right-hand corner. The telephone company, e.g., the BPO, collects the money you owe for using viewdata, and keeps its own part and sends the rest on to the people who provided the information, e.g., the weather bureau.

The computer will usually tell you what your bill is, both for the call you have just made and the cumulative bill since you last paid. In future, it will be possible to set a limit on how much you want to spend, and the computer will stop responding when the limit is reached.

Another important feature is the so-called 'response frame', upon which many of the hopes for viewdata's success are built. Because you, the user, are in two-way communication with the computer, you can answer back to it. The purpose of the response frame is to allow you to place an order or request via the computer. Because you usually only have numbers on your keypad, at the moment at any rate, this has to be done in a rather stylizied way. For expample, you want to order an airline ticket. The viewdata computer tells you to press 9 to place the order. You press 9, and up comes the response frame on the screen. It looks a bit like a printed order form, and it already has on it your name and address (because the computer, as we have already said, knows who you are). Against each choice offered to you—name of airline, starting point, destination, date, flight number, number of tickets, credit-card number—you key in, using the same keypad, the relevant number (e.g., 2 for British Airways, 3 for London, 4 for Paris, 6.8.80 for date, 142 for flight number, 3 for number of tickets, 1234567 for credit-card number). The computer then says, 'key 1 to send, 2 to cancel', giving you the opportunity to change your mind. But if you key 1, that order is then sent off to the airline or agent in question.

If and when people have full alphanumeric keyboards, that is, with letters as well as numbers, then obviously much more complex messages, like electronic letters and memoranda, can be sent by the same means. The recipient is told, when he or she dials in to Prestel, that there are new messages waiting, and can scan them one by one and expedite the order or respond to the letter.

Here viewdata ceases to be a passive information medium and becomes an active functional medium (a theme explored in particular in the International Survey, Part Five) and in many people's opinion this is the most distinctive feature of viewdata, what will in due course be the core of its appeal. But if so, it will challenge all sorts of vested interests, from the postal side of the Post Office (if Prestel becomes a substitute for written messages) to travel agents (if people order travel tickets and holidays direct using viewdata), and many others.

Three basic points can be made about viewdata, points which are elaborated elsewhere. Firstly, viewdata is a 'choice medium' in which the user does not find anything by accident or serendipity. He has to choose to look at something, according to some more or less well-defined purpose. Secondly, it is a colour medium, unlike most computer systems, and that makes it suitable for purposes far outside the reach of those other systems. Thirdly, it has a particular method of working it, the indexes, which are peculiar to viewdata and demand that the user can make decisions (i.e. choices) fairly readily and sensibly.

The technology of viewdata will certainly advance in the future. Already, since the chapter by Keith Clarke (Chapter 2) was written, the British Post Office has announced 'Prestel Two' or 'Picture Prestel', which is a development that allows Prestel to show far better pictures and graphic displays than at present. This will in principle bring Prestel back into the same league as rival systems which have (as is made clear in the International Section) made claims to far superior graphics. Future technical developments could also include hooking viewdata up to existing computers so that the data on them is simply transmitted to the enquirer by the viewdata distribution method: electronic funds transfer by viewdata, for example, to settle bills; local intelligence in the viewdata TV set, attained by putting a micro-computer inside that can perform tasks and search for information for you; and the spread of something that exists already, called telesoftware. Telesoftware is an ingenious idea that takes viewdata to be a means of distributing computer software rather than information as such. The user, with his local intelligent terminal, calls down a program from the viewdata computer to perform a certain task, and when the task is completed he wipes off the program, knowing that it can be called down again at any time from the viewdata computer.

Thus viewdata enters into the field of the personal computer, and it emphasizes the future technical overlap, competition, integration (depending on which way it goes) between lines of development that are at present separate. There are those who say that viewdata will anyway cease to be a distinct development, and before long sink back into the general computer field from which it originally emerged.

Viewdata, for better or worse, is technology driven, like so many electronics industries, and there is no saying where technology will drive it in, say, twenty years' time. But, as the chapter on the television industry (Chapter 5) makes clear, viewdata operates under constraints from other parts of the electronics industry also, and these may have expectations of their own. As the chapter on international standards (Chapter 4) makes clear, viewdata may also find itself in an international market place where diplomacy and technology interact.

But for all that, the technology of viewdata is not in itself unique: its uniqueness lies in the way that various elements have been brought together to produce a certain effect, namely computer information retrieval easy enough for the layman to understand and use.

2. The design of a viewdata system
Keith Clarke

Introduction
Like all successful engineering designs, viewdata has been closely tailored to the requirements of its potential operators and users. Thus it is expedient to examine its general design principles before its technical detail. The system, designed for use by telephone-operating bodies, has three objectives. It must operate as a profitable computer information retrieval service using the usual financial criteria. Cross-subsidy from other telecommunications services is not usually thought desirable and this imposes a tight economic discipline on the designer. Telephone authorities, like most public utilities, suffer from the fact that their traffic tends to be concentrated into certain geographic areas and fairly narrow periods of time. Thus any system that can spread this load, in particular by generating telephone traffic on residential lines at off-peak periods, will improve the economics of the entire system to the benefit of both the operating authority and its customers. Finally, viewdata has obvious potential in the field of message services, and this fact should be fully recognized in any design. The commercial strategy adopted to achieve these objectives should also be taken fully into account. In Britain the Prestel service operates on a tripartite basis. The Post Office operates the computers and writes the software, and also maintains and provides the communications links. The television industry provides TV receivers for the domestic market through its normal retail and rental outlets. Information is provided by information providers (IPs) who maintain the data base. This strategy dictates some of the requirements for the system designer. For example, the decision to leave TV set provision in the hands of organizations outside its direct control meant that the Post Office was confronted with an unusually severe attachment approval problem since the telephone network has to be safeguarded against dangerous voltages and spurious tones, not merely when a new set is commissioned but throughout its life when electronic components may age and the equipment may be subject to indifferent maintenance. The decision to accept a large number of IPs on a common-carrier basis gives rise to the need to handle many simultaneous editors and increases the severity of the accounting and audit problems when information charges are levied.

A viewdata system can be succinctly described as an information and message service for the general public that exploits TV receiver technology, public switched communication networks (predominantly telephone networks at present), and mini-computers in order to achieve the required low costs and simplicity of use. Five major system components can be identified. These are the terminal, the communications link connecting the terminal to the local information-retrieval computer, the local information-retrieval computer hardware, the local computer software, and the computer network by which information is disseminated between the various centres. We will examine each of these in turn, but before doing so will consider the system design philosophy.

System design considerations

Viewdata involves no technology that was not previously known to the TV, communication, or computer industries. This has been cited as a criticism but commercial decision makers usually regard it as an advantage. It is the system design philosophy that distinguishes viewdata from previous computer systems. Two principles predominate. The first is an insistence on regarding a viewdata system as an integrated whole and not simply an interconnected series of terminals, communication links, and computers. The second is an emphasis on simplicity, and a regard for the market intended. The new science of electronic publishing draws on three major older disciplines, broadcast engineering, telecommunications, and computing science.

Let us consider each of the disciplines in turn. Broadcast engineers have an adage that it is better to perform a system function once and well at the transmitter rather than a million times badly in the receivers. This holds good for broadcast systems because the cost of operating the transmitter is not a function of the number of receivers tuned in, but the cost of operating a time-shared computer does vary with the number of terminals connected and with the jobs that they are performing. Thus the viewdata system designer is more likely to incur some costs in the terminal, in order to minimize overall system costs. For example, logic to address individual character positions is provided even though one could theoretically design a system based on the retransmission of a page each time a character alters.

Communication engineers traditionally (and properly) wish to design networks that allow any one terminal to connect to any other and to achieve the maximum degree of interworking between different systems and nations, and this can result in substantial overheads. An example is the number of pins on the CCITT V23 modem (modulator–demodulator) interface. In order to minimize costs for the domestic customer, viewdata terminals are designed to communicate only with the viewdata computer and the computer parts are designed with only viewdata terminals in mind. The use of viewdata terminals on other computers and the use of general-purpose terminals on viewdata is not prohibited by technical or policy barriers, but generally speaking system costs are not incurred to facilitate this. For example, in the UK a 'short-haul' modem is specified on the assumption that Prestel computers will be located near major telephone exchanges, and this means that transmission to computers located elsewhere is not guaranteed.

The necessity for this discipline becomes apparent when one considers the candidates for compatibility with viewdata. These include electronic mail, teletext, word processing, Telex, main frame computers, packet switching, VDUs and the typewriters and keyboards used throughout the world. Obviously compatibility is a desirable objective, but the cost of a terminal bearing the overheads associated with each service would be prohibitive.

Some computer professionals tend to eschew simplicity of operation in the search for elegant information-retrieval techniques. Keyword searches using thesauri and full alphanumeric keyboards are fine for professional use, and indeed it is not claimed that viewdata is an alternative to the specialized services now in operation, but they are not suitable for the average customer at home.

For all the above reasons the orthodox wisdom of the communications computer and broadcasting professions does need to be questioned occassionally when videotex standards are discussed.

Viewdata terminals

Figure 2.1 is a schematic diagram of a TV receiver capable of receiving both viewdata and teletext. The heart of such a receiver is the memory. This stores information received via other communications media in digital form ready for display on the TV screen. Information is extracted from the memory by the display circuits and fed to the three colour guns (assuming it is a colour receiver) of the set. The chief component of the display circuits is a device called a character generator, which converts the digital information into the series of dots (bright or blank) which comprise the characters to be displayed.

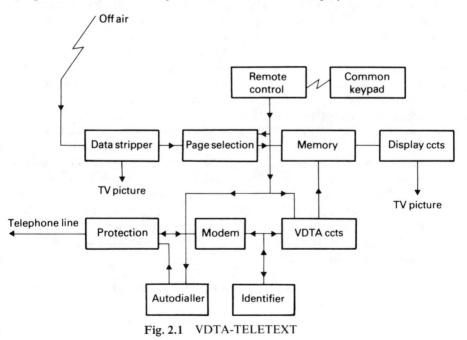

Fig. 2.1 VDTA-TELETEXT

Let us consider first a customer receiving information via broadcast teletext. He or she enters the page number required from the keypad. This will almost certainly be a remote-control keypad which communicates with the set electronics via either an infra-red or ultrasonic link. The page number required is entered into the page-selection electronics. The television signal is received off air and passes through an integrated circuit known as the data stripper. This separates the analogue TV component of the signal from the data components. The pages are broadcast in rotation. The address of each page as it is received off air is compared with the address of the required page by the page-selection electronics. When a match occurs, subsequent page data is entered into the memory.

Now consider the receipt of data from the telephone line. The set is permanently connected to the line via a plug and socket, and thus protection is most important. There are within a TV set several sources of voltage which must be isolated from the telephone network. This network is essentially a low-voltage system and the protection of it, and its maintenance engineers and customers, is of prime importance. Dangerous voltages occur in the mains power unit, in the extra-high-tension supply, in the focusing coils, and could also occur because of lightning-induced surges on the aerial. Thus the provision of adequate protection by isolating transformers and/or fuse barriers is essential. Call set-up is performed by autodialler circuits which are initiated from the keypad. Data is received from the telephone line in analogue form and the modem (modulator–demodulator) is necessary to convert this data to digital form. For the UK Prestel service this modem is usually integral with the TV set, although other administrations regard this as a monopoly provision item which must be provided separately outside the set. Once the call is established, the set must be identified to the computer for billing purposes. This function is performed by a device known as a hardware terminal identifier, which transmits a unique series of numbers on receipt of the enquiry code that is transmitted at the start of every viewdata call. Data received from the telephone line is inserted into the memory after passing through the viewdata control circuits. These have two main functions which are slightly different from those of the data stripper on the broadcast teletext side of the circuit. They convert the data from the bit serial mode in which it has been received to the 7-bit parallel mode that is used within the memory and display circuits. They also separate the data characters that are to be displayed from certain control characters which format the data. Typical of these are back space, horizontal tab, line feed, and vertical tab. Such characters are not displayed but alter the address in memory (and hence on the screen) of subsequent data. In modern sets this circuitry will be reduced to one or two large-scale integrated circuits.

The communications link between the terminal and the local computer

Figure 2.2 illustrates the communications media that could be used. Let us consider their various merits. Broad bandwidth links, either coaxial cable or fibre optics, have undoubtedly a substantial future in the field but are precluded from widespread use at present simply because it will be some time before the cities of Britain can be rewired. Audio tapes, video tapes, and video discs will all find a

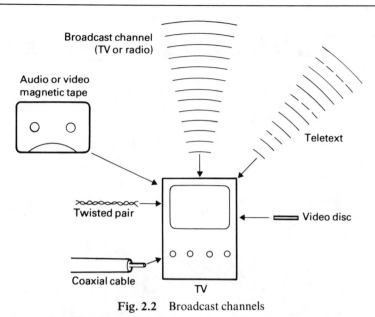

Fig. 2.2 Broadcast channels

role for the storage of information that is to be displayed on television sets but are not suitable media for prime distribution because any such method would be labour intensive. The provision of a broadcast channel, either TV or radio, for data purposes would be cheap but would suffer from two major disadvantages. The first is that such channels are in demand for conventional TV and radio. The second is that they do not provide a return path between the customer and the computer; they are one-way only. The insertion of data signals into the field-blanking period of existing TV broadcasts, as is done for broadcast teletext, is cheap and effective for small volumes of data (since most of the channel information capacity has already been utilized for TV purposes) and again there is the lack of a return path from the customer to the computer. This leaves us with the conventional 'twisted pair' of copper wires which constitutes a conventional telephone line. They are numerous and in many instances under-utilized and they provide the required return path to the customer. Digital data is transmitted by switching between two analogue tones. The system works in asymmetric full duplex mode, which means that both send and receive channels are working simultaneously but at different speeds. Data is transmitted from the computer to the terminal by switching between two signals of 1300 and 2100 Hz at a rate of 1200 bits per second. Data is transmitted from the terminal to the computer by switching between two frequencies of 390 and 450 Hz at a rate of 75 bits per second. The much lower frequency on the return channel is tolerable because it approximates well to the maximum rate at which data can be generated by a keyboard operator. Data is transmitted in asynchronous form, which means that there can be gaps in the transmission when no information is being sent. The data is sent in the form of 10-bit characters according to international standard

Bits b7 b6 b5 →	0 0 0	0 0 1	0 1 0	0 1 1	1 0 0	1 0 1	1 1 0	1 1 1
Col / Row (b4 b3 b2 b1)	0	1	2 / 2a	3 / 3a	4 / 4a	5 / 5a	6 / 6a	7 / 7a
0 (0000)	NUL		—	0	@	P	|	p
1 (0001)		Cursor On DC1	!	1	A Alpha" Red	Q Graphics Red	a	q
2 (0010)			"	2	B Alpha" Green	R Graphics Green	b	r
3 (0011)			£	3	C Alpha" Yellow	S Graphics Yellow	c	s
4 (0100)		Cursor Off DC4	$	4	D Alpha" Blue	T Graphics Blue	d	t
5 (0101)	ENQ		%	5	E Alpha" Magenta	U Graphics Magenta	e	u
6 (0110)			&	6	F Alpha" Cyan	V Graphics Cyan	f	v
7 (0111)			'	7	G Alpha" White	W Graphics White	g	w
8 (1000)	Cursor ← BS	CAN	(	8	H Flash	X Conceal Display	h	x
9 (1001)	Cursor → HT		)	9	I Steady	Y Contig Graphics	i	y
10 (1010)	Cursor ↓ LF		*	:	J	Z Separated Graphics	j	z
11 (1011)	Cursor ↑ VT	ESC	+	;	K	↓	k	¼
12 (1100)	Cursor Home & Clear FF		,	<	L Normal Height	½ Black Background	l	||
13 (1101)	Cursor ← CR		-	=	M Double Height	↑ New Background	m	¾
14 (1110)		Cursor Home RS	.	>	N	← Hold Graphics	n	÷
15 (1111)			/	?	O	‡ Release Graphics	o	█

(Columns 2a, 3a, 6a and 7a contain graphic block characters.)

Fig 2.3 Viewdata transmission codes

38

ISO 646. There is a start bit, this is followed by seven data bits, a parity bit which acts as a check against errors, and a stop bit. Figure 2.3 shows the codes that are used. In their simplest form the 7 bits give rise to 128 code combinations which are represented by 8 columns, each of 16 codes. This repertoire is extended by using an escape character which enables certain codes to have two meanings. When considering the viewdata transmission scheme it is important to remember that the cost of the modem, which converts the analogue signals to digital signals and vice versa within the TV set, is a function of the maximum loss likely to be encountered between the set and the local computer. To minimize this cost, Post Office system designers have assumed that the local computers will be located close to the major switching nodes of the telephone network. These are known as group switching centres and are usually located in the main towns and cities. Thus in order to achieve cost minimization a more limited data transmission scheme than that used by the coupled Datel services is utilized.

The choice of a computer

The computer used for viewdata had to meet a number of objectives. Customers on viewdata and on other time-sharing computers access the computer by what are known as input/output ports. The capital cost associated with each input/output port must be kept to a minimum. However, in achieving this the number of ports must be kept to a level that telephone network planners can provide within their normal planning cycles. It would not be practical to minimize the cost per port by providing a huge number of ports on a large expensive computer because telephone lines could not then be readily provided. Accomodation costs must be kept to a minimum. Traditional computers require false floors, rigorous air conditioning and strict dust filtration. These costs often account for one-third of the cost of running the computer, and savings in this area should be sought. The viewdata system designer seeks ready access not only to the software but also to the hardware provided by the computer manufacturer because he will be writing his own operating system, and may wish to connect his own hardware in the form of front-end processors. He will require fast interrupt handling to facilitate the rapid input and output of data, and because his information will be stored on rotating magnetic discs he will also require fact disc access. Because viewdata applications give rise to much input and output, he must be provided with either a large amount of store for buffering purposes or he must use front-end processors. These criteria push the system designer in the direction of real time, communication or mini-processors rather than large data-processing main frames. The PO Prestel service uses a computer known as the GEC 4080 mark II. The dominant characteristic of this computer is a microprocessor known as Nucleus which performs by hardware some of the functions performed by software in previous generations of computer. In particular, it:

- enforces protection between the boundaries of processes (programs) to ensure that they cannot interfere with each other in an uncontrolled fashion
- provides safe channels of communication between different processes and between processes and input/output channels

39

- performs short-term scheduling for the computer by ensuring that at any time the most urgent process which has useful work to do is in control of the central processor.

To perform these tasks, Nucleus maintains system tables which define for every process the areas (segments) of main store that a process may access and the other processes with which communication can be established by means of inter-process messages.

The viewdata software

This is best described as it was developed. The foundation of the viewdata software is the question-and-answer technique that was developed to ensure ease of use. A simple question-and-answer technique is employed as the basis of the access method and the slogan 'a prompt on every page' has been adopted.

In addition, a limited repertoire of 'star' commands has been devised to perform additional functions. These are listed in Table 2.1. The basic software to perform information retrieval using the tree method and the star commands is described in the section on the viewdata pilot trial system. This was known as the 'pilot-trial software'. Later a number of utilities were added to form the public-service software and these are outlined in the section on operational software.

Table 2.1

	Command		Implementation
1	User No		Automatic Entry
	*0	2	Go to Start
	Jump to page 'n'		*n
4	Back one page		*
5	Delete keyed entry		* *
6	Repeat page from disc		*09
7	Repeat page from store (used as a check against errors)		*00

The viewdata pilot trial system

The scheduling which the GEC 4080 does by micro-program is by process priority. To write a viewdata process and replicate it 200 times would have required scheduling by time slice to give every user equal priority. This would have had to be done by software and would have thrown away a major advantage of the computer. So it was decided to write viewdata as a small number of processes, each exececuting a well-defined task and doing it for all the users in turn. These processes could then be sensibly scheduled on a priority basis. In essence, one more step was taken towards the process-control type of structure and away from conventional time-sharing. This design has the theoretical disadvantage of increasing the system overhead in inter-process communication but that is handled by the machine's micro-programmed executive and so the overhead is greatly reduced.

Thus a design was chosen in which the main system overheads were executed by the hardware executive or by highly optimized code. From this and the brevity of the code required to give the user the simple facilities offered, it was apparent that the system would not be processor-bound.

The data structure

The accent is on simplicity of use for the general public. It was felt that simplicity also implied generality, so that all parts of the database should be accessed in the same manner. The method chosen is to offer data selection by a choice from a list of up to ten selections. Where the classification of the information requires more selectively than that, further levels of up to ten choices are offered. Eight such levels are available if necessary to select the required information. The reason for the choices being restricted to ten is so that, with a simple keypad each button depression will elicit a response, so avoiding an unsophisticated user failing to rouse the system by mis-keying a character string. Although the system as written has facilities to input character strings, use is not made of this in the main data structure in case it should prove unacceptable to the public when they are first introduced to the system.

The data structure indicated by the above is a *tree structure* as shown in Fig. 2.4. The nodes of the tree are called *pages* to emphasize the similarity with the pages of the teletext system. Both are blocks of information identified and selected by a string of digits, the page number. Obviously our page numbers are dictated by the page's position as a node in the tree. Like teletext, our pages consist of one or more *frames* and where more than one is present they are displayed one after the other. Unlike teletext, our second and subsequent frames are displayed at the instigation of the viewer who can therefore proceed through the information at his own pace and not that dictated by the service. And he will also always come in at the first frame, so preventing, we should guess, some confusion. Where a page presents a viewer with a selection of subsequent ones, the viewer may make the selection at any frame of the page and not just the one which displays the choice to him. This is to allow for the user's anticipation as he becomes familiar with the information. A user can anticipate as many levels as he likes and as long as he has done it correctly he will arrive at the required page. Generality has been retained wherever it does not affect the efficiency of the system.

Accessing the disc

Since the general public is unlikely to be as tolerant of slow response as is the professional user, disc traffic has been kept to an absolute minimum. Mathematical modelling has shown that even with inexpensive discs a response time as low as a second or so can be achieved if disc-queue optimization is used,

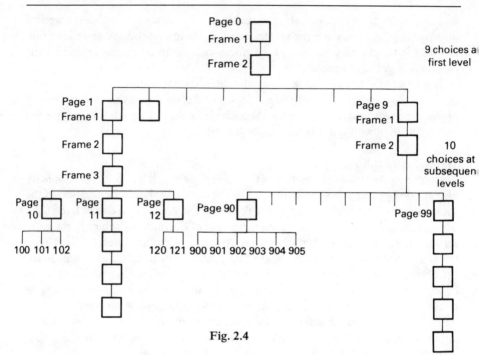

Fig. 2.4

provided that only one disc request is generated per user request. Obviously this proviso could be weakened by using faster discs and several independent disc channels, but we decided to keep such methods in reserve (mathematical models have been known to give optimistic results) and accept the restriction of one disc request per user request, at least in the great majority of cases.

This meant in the first place a core-resident system to prevent the automatic overlaying of segments on to disc by the machine's virtual memory system. This in turn brought us up against a core limit. Each user requires an output buffer of 960 bytes, since this is the size of the display on his terminal and we are allowed only one disc access to fill that display. The maximum core size of the GEC 4080 used for the pilot trial is 256 kbytes, which is sufficient for 200 buffers and the system, but leaves little over for database directories. As we did not wish to limit the directory size, which would in turn have limited the size of the database, we concluded that the directories would have to be disc-resident. So the system is not totally core-resident but all the code and the users' buffers and other data areas are.

The problem then resolved itself into one of minimizing the times that a user needed to access a directory when requesting a display (or 'frame'). It had already been decided that for simplicity in use by the general public a tree structure should be used for the database, and it is a feature of such a structure that each frame 'knows' the choice of frames open to the user from itself. Thus it can have attached to it a table of such choices and, by referring to the directory when constructing the database, such a table can be one of real disc addresses. At the cost of more directory accesses when setting up or modifying the database, the need for directory access when the general user reads from it can be eliminated. The system is envisaged as one in which vastly more work will be generated by general public access to read the database than by access required to modify and update it.

In fact, the table of choices attached to each frame allows those choices to be either real disc addresses or page numbers which require dictionary access. This allows a less rigid data structure and, more importantly, allows references to pages to be entered before the pages themselves. It is not envisaged that significant amounts of the database should be outside the strict tree structure, as that would lead to an increase in disc traffic and deterioration in response.

To sum up, the whole system of computer, software, and database has been chosen to perform simple operations at high speed so that the general public can be offered, not a computer system, but an information system at an attractive price.

System structure
The system is table-driven. It consists of the following processes:
INPUT process
OUTPUT process
GATE-keeper process
DISC-handler process
TASK processes
Communications between processes is by messages and by shared segments defined by these messages.

Figure 2.5 is a diagrammatic representation of the viewdata system. In it a

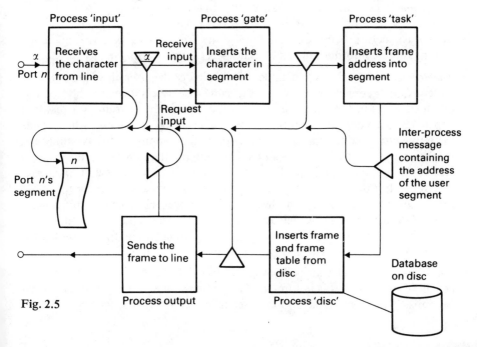

Fig. 2.5

user can be considered as a message that is handed round the system. When a user inputs a character, INPUT places it in a message identified by the user's port number and sends it to GATE. After checking that the system is ready to receive it, GATE sends the message with the input character to the appropriate TASK. How GATE chooses the appropriate TASK will be described later. The TASK, using the input character and the data on the user's state which is passed to it in the message, selects the description of the next frame to be output. It sends this via the user message to DISC which obtains the frame from disc and places it in the user message. It then sends the message to OUTPUT to transmit the frame to the user. OUTPUT then sends the user message to GATE to await the arrival of the next character from that user.

Other paths between processes are possible, and will be described later. The above is, however, the simplest and most common.

The driving tables in the user segment
Each user has two tables which, with buffers, data area, and the space for the information to be displayed, fully define the user's state. All these areas are held in the *user's segment* which is the item sent round the system, and are as follows:
   The frame
   The frame table
   The user table
   The input buffer
   The user work area
The areas are listed above in the order in which they are set in the user's segment. Further details are given below.

The frame. The frame is a region of 960 bytes which receives from disc the information to be output to the user. The contents of this region may be modified by a Task before being output, but in any case it is a passive region for transmission of information and does not steer the processing in any way.

The frame table. This is the table which directs the user through the data structure. It is copied, by DISC, in the same disc access as when the frame is obtained and must be contiguous with the frame. Every frame on disc has appended to it a copy of the frame table which is relevant to a user when he has the frame displayed to him. It governs entirely the response of the system to the user's reply to the frame. It does this by specifying the number of the Task to receive the user's reply and the list of frames which can be accessed by a user at this frame. It is the Task's job to receive the user's input and apply its algorithm to it to come to a selection from the list. It then moves this selection into the first entry of the list where DISC will access it to find the frame next required.

The frame table also contains the frame's page number and frame number for identification purposes. Its full layout is shown in Fig. 2.6.

The user table. This contains all of the interface data not held in the frame table, i.e., all of the data that is not updated every time a new frame is brought down from disc. The frame table and the user table define the interface between the

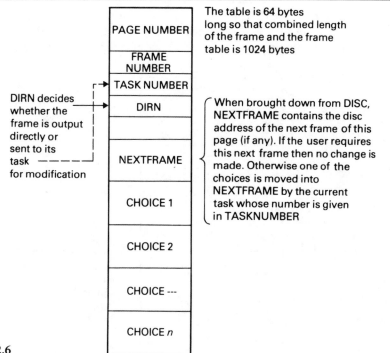

Fig. 2.6

processes GATE, TASK, DISC, and OUTPUT. The structure of the table is shown in Fig. 2.7.

The cyclic history buffer is there to provide the 'go back one frame' facility. This facility cannot be provided using the data structure, since the user is allowed to jump at random about the structure. The main purpose of the facility is to allow a user to correct an erroneous choice by going back a few frames to make the choice again. Consequently the buffer is small (four entries). The 'go back one frame' command will cause the current frame pointer to be stepped back one frame cyclically. The end of buffer pointer will not be changed. If stepping back the current frame pointer would bring it back into coincidence with the end of buffer pointer, the step back does not take place and the command has no effect.

A frame address is put in the buffer by DISC when it obtains the frame. The buffer is written to by DISC and read from by some TASKs. PORTNUMBER is used by OUTPUT to find the line down which it must send the frame.

SENT is set by GATE when it sends a user segment into the system. If more input is received while SENT is set, GATE queues the new characters in the input buffer and sets RECALL. When OUTPUT receives a user segment it will not, if RECALL is set, initiate the output of the frame before returning the user segment to GATE. RECALL is in fact a count of the contents of the input buffer used by GATE, but by sharing it with OUTPUT we hope to save on some peripheral traffic and give the user a less erratic-looking display, by eliminating the output of the intermediate frames that he is not interested in.

STATE is a byte used to inform a TASK if a message is bringing a single character from GATE, or a component of a character string from GATE. CHAR is a byte containing the next character of input on which the system is going to act.

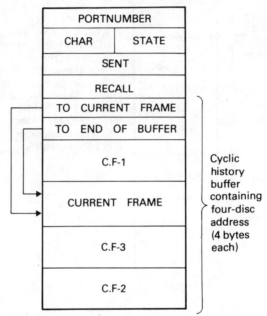

Fig. 2.7

The input buffer. Input may be received by GATE while the user segment is traversing the viewdata system. In that case the input must be buffered. The input buffer is a data area private to GATE in which this buffering takes place in a cyclic buffer of bytes.

The user work area. The last area in the user's segment (last to allow it to be extended if necessary) is a work-space area available to TASK which uses up the tail end of the segment which would otherwise be wasted in fragmentation, as all segments must be multiples of 64 bytes. This area can be used to allow intermediate results to be stored, as when a TASK is assembling a value from a stream of input.

Outlines of the actions of the system

This section identifies the routes that the user's segment may take in traversing the viewdata system.

The simple route (see Fig 2.8)

    I.   Input a character to the user table by INPUT and GATE

   T.   Send user's segment to the appropriate TASK. This analyses the response and specifies a new frame by filling the next frame entry in the frame table.

   D.   The required frame is obtained from disc by DISC and the frame table in the user segment is updated.

O. The frame is output and the user's segment is returned for another character.

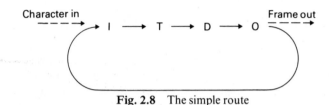

Fig. 2.8   The simple route

The simple route with additional input

Here a TASK may require more than one character before it is in a position to send to DISC. Since the TASK will probably be storing intermediate results in the user work area, input has to come in under a different state so that the task is aware of this. The type of input TASK is defined by STATE which is set to *b* by a TASK to get additional input for itself. Since no TASK can specify the value of STATE for another, it must set STATE to *a*, the normal value before sending the user segment to DISC. So far the route above, the transitions have the value of STATE shown in Fig. 2.9.

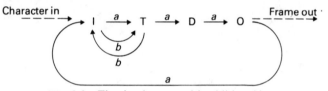

Fig. 2.9   The simple route with additional input

The modified frame route (see Fig. 2.10)

A frame table may specify that the frame obtained by DISC is to be sent to its TASK for modification before being output, e.g., by having today's date inserted.

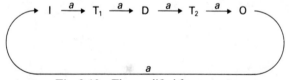

Fig. 2.10   The modified frame route

The multi-task route (see Fig. 2.11)

A TASK may decide, after some time, to relinquish control to some other task which it specifies by changing the value in TASK NUMBER and sending the user back to GATE. This is done by the command process when it abandons a command and restores the user to his original TASK.

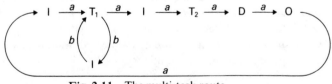

Fig. 2.11   The multi-task route

The viewdata processes

These are the processes which implement the system structure outlined above.

INPUT

This is our special-purpose line driver for the incoming (75 baud) data lines. It communicates in the most efficient manner with the input/output processor handling these lines and acts on the incoming data one character at a time. Since the incoming line is responsible for handling the modem of the corresponding port, INPUT also detects when a line is down, in which event it will reset the port's user segment to an initial state ready to receive the next call. Other line errors are also handled by this process.

There are in fact not one input but thirteen or so. This is because the hardware microprogram presents the interrupt information in a form requiring less software processing if only sixteen lines are connected to each process. Each input communicates with GATE in an identical manner and GATE is not aware how many there are. Each driver handles sixteen ports and owns the user segments corresponding to each port that it handles. As input is received, INPUT will hand the corresponding user segment to GATE, as shown in Fig. 2.12.

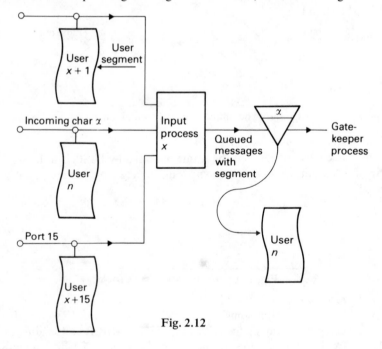

Fig. 2.12

GATE

The task of this process is to accept characters, one at a time, from process INPUT and feed them, one at a time, into the viewdata system. It only feeds them into the viewdata system when that system has passed it the user segment to indicate that all previous work for that user has ceased. In that way the input to the system is handled in an orderly manner.

Since the rate of input by the user will not be equal to the rate of acceptance by the viewdata system, GATE must buffer the incoming characters if they come in too fast, and must have a way of storing and retrieving the user segment if they do not come in fast enough to keep up with the viewdata system. It achieves both these objectives by being sent the user segment by its *owner*, the process INPUT.

If the user segment is simultaneously being passed round the viewdata system, GATE accesses it, to buffer the input from the message into the area accessible only to itself. GATE then relinquishes all knowledge of the segment until presented with it again by INPUT or the viewdata system. When the system returns it, GATE will refill the current character entry CHAR from its buffer and re-enter it into the system. However, if on this return the buffer is empty, GATE again relinquishes all knowledge of the segment. The user segments of quiescent lines will then automatically 'queue up' in the INPUTs and will automatically be sent to GATE again when there is more input.

TASK

These are processes called in by frames to deal with the user's response to the information displayed by the frames. Their detailed coding is therefore to be decided by the designer of the data structure that they serve. They do, however, have standard subroutines to implement the interface with the rest of the viewdata system.

At present three tasks are written. Briefly, they are:

The command task. There is a simple command language in which all commands start with the warning character '*'. GATE intercepts all '*' input and sends them to the command task regardless of the value in the TASKNUMBER word. The command task then sets TASKNUMBER to its own value to receive and act on the rest of the command. If however, on receiving a '*' it finds that the task number is already its own, it interprets this as a cancellation of the command, restores the original task number and sends the user back to GATE to await further input.

The frame getter. This accepts a number in the range 0 to 9, and ♯. If the character is '♯', it sends the user segment directly to DISC to get down the next frame of the current frame, as defined in NEXTFRAME of the frame table. A number is used as an index to select one of the choice entries to be moved into NEXTFRAME. When the move has occurred, the user segment is sent to DISC.

The error task. If a frame or page does not exist, an error frame is output. This frame's frame table contains the task number of the error task so that the next input of the user is handled by the error task. Regardless of what that input is, the error task consults the HISTORY vector and loads the address of the previous frame into NEXTFRAME and sends to disc so that the user has the opportunity of correcting his choice.

DISC

The primary purpose of this process is to obtain frames from the disc store. It interfaces directly to the disc drivers and does not use the disc filing system, so avoiding the overheads associated with that system. This process receives the user segment from a TASK when that TASK has selected the frame to be displayed next. The frame which is required for OUTPUT is held in NEXTFRAME. A positive value is a real disc address, a negative value indicates a directory access to find the real address, and a zero means the last response made was illegal; in this last case an error frame is displayed, the response being sent to the error task. When the real disc address of a frame has been obtained, either directly or via the directory, the user's history table is updated. A message is then sent to the disc driver to obtain the frame. In due course a reply is sent the frame buffer having been filled, and DISC then decides where to send the frame. If the variable DIRN is zero, the user's segment is sent to OUTPUT. Otherwise the segment is sent to the TASK specified in TASKNUMBER.

OUTPUT

This is a special line driver for handling the outgoing (1200 baud) data lines. It displays a whole frame, preceded by four control characters to clear the screen, in a single autonomous transfer. If the user interrupts during output, the next frame is obtained from DISC, the current transfer is aborted, and the new one is initiated. As with the INPUT processes, these drivers can handle sixteen lines most efficiently and therefore there are about thirteen OUTPUT processes, each handling sixteen users.

The database processes

So far we have considered the system only as supplying data to the general user. Since most of the traffic is to be in this field, this is where the optimization has been done and it is where most of the features lie. However this optimization has led to an unusual type of database, and this in turn is handled by three subsidiary, but relevant, areas of the system. These are:

DINT—the disc initializer

DIdd—the processes which create the frame and frame table

UPDATA—a procedure in DISC which inserts new entries in the database when it receives them from the processes DIdd.

Only UPDATA is strictly part of the viewdata system. DINT is run under ordinary DOS 2 operating systems to initialize a disc into the correct format for use by the system. The processes DIdd could be in another computer if required. They run asynchronously with viewdata and their interface is to send 1 kbyte of data to DISC from time to time.

The disc initializer process

This divides the disc into 1 kbyte blocks. It creates two kinds of block; system blocks which are known to the viewdata system by their disc addresses and free data blocks which are known to the viewdata system by being chained together in the free chain, which is itself located by head and tail pointer in a system block.

The initializer creates a system region of three areas (Fig. 2.13). The first area consists of one block which contains in its first two full words the addresses of the first and last blocks on the free chain. The second area contains blocks to be filled by error frames. The third area contains the directory blocks which will contain the disc addresses of all the pages of data held on the disc when it is in use. These directory blocks are initialized to zero by DINT.

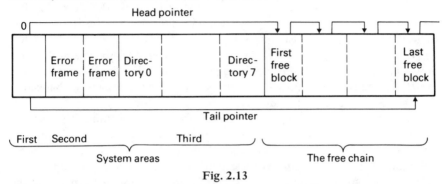

Fig. 2.13

The data input programs
Each data-supplying station is connected to one such program tailored to that station's needs and restraints. These programs construct the frame and frametable and send them in a 1 kbyte segment to the process DISC to be included in the data structure by the procedure UPDATA.

The interface via FRAMETABLE is protected by UPDATA to the extent of checking for and ignoring segments with impermissible data which if accepted would crash the system. It does not, however, check for incorrect data that would corrupt the database. It is, therefore, the responsibility of the data input programs to ensure that only sensible data is entered into FRAMETABLE.

The data assembler procedure
Data is presented to DISC on a higher priority route than the data requests from viewdata, so that data updating is not shut out by heavy user traffic. There is a great danger when updating a disc data store that a system failure during it will leave the data structure in an inconsistent form and thus cause the loss of the whole data store. Care has been taken to write the procedure UPDATA in such a way that no such inconsistencies can occur.

The main operation to protect the data structure is the method of chaining the data blocks. Blocks are moved from the free chain to a page's chain of frames and vice versa, and this is done in such a way that the chain pointers are never inconsistent. The data updating process holds a linked list of the 1 kbyte disc regions that are not filled with current information, i.e., that are free. It adds to and takes from this list in a first in–first out sequence. It therefore takes the oldest such record and copies the new frame and frame table into it.

The filling of the choices in the frame table must now be done. The system is optimized to a strict tree search, that is, where the response 9 to page nnnn gives page nnnn9. This is achieved by all such choices being translated into real disc addresses and all others being left as the page number as supplied by the data input program. If the page below does not exist, the choice is left as a page number. (When the page below is finally produced, the parent page is updated.)

Next this new frame must be connected to the data structure. If the frame is other than the first of a page, the one it is replacing is obtained (if it exists) and the disc address in its NEXTFRAME copied into that of the new frame. Then the frame just above the one being replaced is obtained and its NEXTFRAME has the disc address of the new frame copied into it. If the frame is the first of a page, then the page directory is searched to find if it yet exists. If it does not, then a new entry is put in. If it does, its NEXTFRAME is entered as above and the page directory entry altered. If the page is page nnnx, then the xth choice is found in nnn. Where the choice is a real disc address or page nnnx, its value is changed to the disc address of the new frame in each frame of page nnn.

Whenever a frame is entered in the data structure, the frame it replaces is placed at the end of the free queue. Any user whose user segment has the address of this frame will still be able to access it. If the user accesses the frame within a reasonable time, he will still be presented with valid, if not quite up-to-date information. Otherwise he may find himself transported somewhere else in the data structure. He will always get some response and will not break the system.

When the process DISC is first set running, the routine initialize is called. This routine reads in the pointers to the head and tail of the free queue and the directories from disc. Control is then returned to the main code of DISC until a message is received from a data inputting process, when UPDATA is called again.

Operational software

The preceding sections have described the original, simple viewdata system. To run an operational public viewdata service, many other functions are required and some 60 000 words of code have been written. The most important additional functions are described below.

Editor. Getting information into to a viewdata system economically and swiftly is as important to the operator as getting it out and displaying it. An editor program to enable information providers to compose data on the screen has been provided. This enables editors to set up the links between pages using simple numeric references and without using absolute disc addresses. Facilities to compose large alphanumeric figures using the simple graphics mosaic elements are also provided. Once created, frames may be amended and deleted simply.

Closed user groups. This provides the information provider with facilities to put users into or remove users from, the closed user groups which he owns. Once in a closed user group, the customer can then view information which is not available to users outside the closed user group. The facility has obvious applications for business use.

Messages. This facility enables customers to create messages that are placed into a 'mail box' area of store and can then be examined by other explicity named users.

Customer billing and IP revenue. In the UK Prestel system the customer pays a time-based charge to the operators of Prestel for the use of the system. He also

52

pays a frame-based charge, which may range between 0 and 50p to the information provider as a royalty. This charge is calculated and collected by the Prestel service so the software must calculate composite bills for the customer and also keep track of the revenue due to be repaid to the information provider.

Optional passwords. Customers may opt to increase their own security by choosing an additional password of their own to be used when they log in to the system.

On-line error-reporting processes. With a large number of processes running concurrently, the possibility of any one process entering an error state has to be recognized and a special error-reporting routine exists to bring such occurrences to the attention of the system operators.

System manager facilities. This process is responsible for the orderly start up and close down of the system and provides the operator with facilities for restricting access to the system to certain classes of user (e.g., information providers only, system manager and/or system editor only).

Bulk update facilities. Facilities are provided for the database to be updated locally from magnetic tape or remotely from a data transmission link to an information provider's computer.

Utility processes. About five facilities are provided to assist in the task of national database management. These include archive facilities for data security and facilities for inserting, replacing, or deleting database frames on an *ad hoc* basis when required to recover from hardware failures or when new computer centres are opened.

Statistics-collecting routines. Information on the use of the database is of crucial importance to operating staff, to information providers, and to the designers for future viewdata systems. Thus an analysis of the type of use made of the system (in a way that is comensurate with data privacy) is required and has been provided.

The design of a national network for Prestel

The first stage of the Prestel public service network is shown in Fig. 2.14. All the information providers access one national update centre for editing purposes. Customers access regional computers (marked as RC on the diagram) and these are usually provided in pairs to increase data security. Thus the network is star-shaped and the database is replicated in each location where service is provided. This simple design of network has some advantages and some limitations. An analysis of the performance limitations of viewdata computers shows that the ability of the magnetic discs to handle frame retrieval is a limiting factor on the throughput of the overall system. Thus additional discs must be provided to yield the required data throughput, and it makes sense to replicate these discs near to the customer rather than at one central location. The fact that data is replicated on a number of sites also means that a high degree of security against catastrophes is achieved. In the long term a star network of this type is not likely to be the optimum. The need will arise for the provision of local data which is not required nationally and local editing facilities will also be required. Moreover,

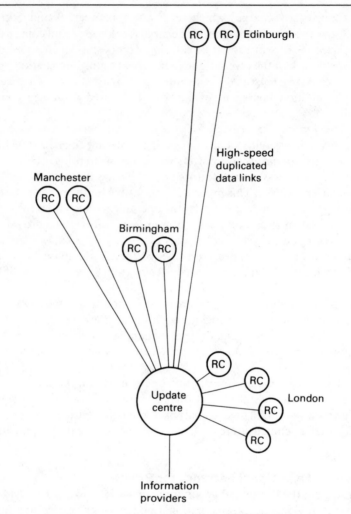

Fig. 2.14 Prestel public-service network first stage

the implementation of the message facility on a national basis will give rise to traffic patterns that cannot be readily handled through a central node, and the gathering to a central location of response frames, which are limited fixed-format messages intended for information providers, will also create problems. Thus a more sophisticated network will be required but it would be imprudent to design such a network until accurate information on traffic flow on a national basis is available. Such information is being gathered by the UK Prestel service and will be used as the data input to a mathematical model to determine the optimum long-term network strategy.

System summary
Figure 2.15 summarizes a number of crucial parameters pictorially.

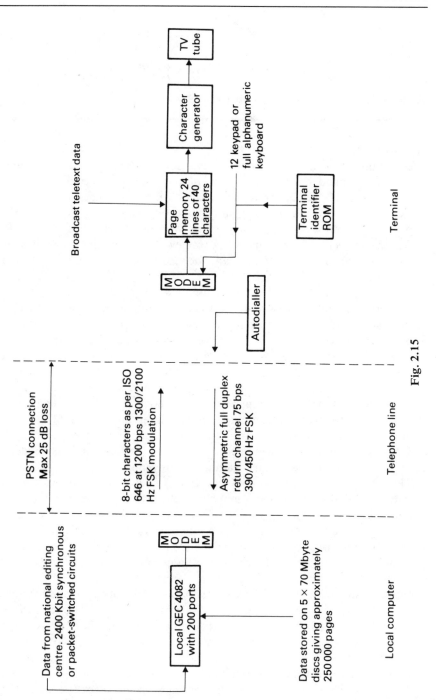

Fig. 2.15

3. Future technical developments

Robbie Hill, Paul Radcliffe and Tim Sedman

Introduction

Prestel is only the beginning. The service is already useful and it is certainly going to grow and expand into something more powerful affecting, in due course, most of the people in the UK either in their business or in their domestic lives.

Yet it is a primitive system. Like many pioneering endeavours its importance is two-fold: first, obviously, in the function or service provided; second, in the fact that it provides a new vocabulary, a new framework for thinking about ideas and yet more powerful systems. Historically the second aspect could prove the most significant.

In their separate ways, technical developments in microprocessors, communications, information retrieval systems, and electronic computer peripherals have been rapid. Prestel has brought many of these threads together in a new synthesis. It is possible to see how, technically and commercially, computer networks and services are likely to be developed, on a hitherto unimaginable scale, to reach inside 20 years a majority of the population of the Western developed countries.

This development means literally millions of terminals and the exploitation of every advanced type of communications technology now being installed, from packet switching networks to satellites, optical fibres, and the 'wired city' of the future. The enormous variety of applications involved, and hence the demanding range of transactions, will have to be supported by terminals and man–machine dialogues, still, for the most part, waiting to be invented.

Three factors are of crucial importance to the success of videotex: terminal design; information range and quality; and effective exploitation of advanced communication technology.

Mass utilization of computer terminals depends on very flexible and low-cost devices. This is discussed in the following section with emphasis on intelligent terminals and the use of telesoftware techniques to extend the scope of the hardware.

Information providers must find it easy to test the new medium and have cost-effective tools available for maintaining a production database. Subsequent sections deal with: editing problems and developments, with the emphasis on the practical lessons learned from Prestel; the symbiosis between videotex and the new developments in communications; and one example of a business market, the office. This illustrates the potential value of videotex as an easy-to-use technique for the uninitiated to handle data files; the problem of compatibility between devices on a single local system; and not least, it is an example of a major market for videotex services and hardware.

In the final section we return to the overall growth in videotex applications and underline the problems, mostly techno-economic, that must be overcome before the 'information society' can become a reality.

Terminals

Types of user terminal

The specification of the early viewdata receivers in the UK required that they should be capable of TV reception, and that they should contain specially designed circuitry. This meant that they tended to be made only by the TV manufacturers. However, the ending of the requirement to receive broadcast transmission, and the increasing availability of standardized 'chip sets' mean that the range of potential suppliers has increased enormously. It now includes communications companies, suppliers of hobbyist computers, terminal manufacturers, office equipment or small business computer companies, and a number of entrepreneurs who have started up specialist companies in this market place.

From this wide range of suppliers, we can see a spectrum of terminals emerging with differing capabilities. The rapidly declining cost of microelectronics means that extra capabilities can be built in for relatively little extra money, and future terminals will become increasingly intelligent, i.e., having some processing capability as well as being simply for displaying data. Although many designs are under development, clear trends are emerging which should lead to ranges of terminals.

The simplest enhancements are pre-programmed into the terminals, and can include the ability to select pre-defined pages automatically, or to store a number of pages in local store for off-line display. Alternatively, the micro can be used to restrict the user to certain types of pages, for example to prevent the use of response frames from public-access terminals. Further developments of this idea include the use of different passwords by different classes of user who are permitted to access different facilities. For example, some may only use a private viewdata system while others can call up Prestel.

A more flexible terminal will have the ability to execute more than one program. With domestic terminals, for example, it will be possible to replace a cassette or a plug-in ROM (read-only memory) and the terminal will be able to offer a range of games or simple educational programs as well as being a Prestel terminal.

If, in addition, the terminal has some RAM (read-and-write memory), it is capable of being a reprogrammable device, i.e., a true computer, capable of a wide range of applications. Here again there is a range of possibilities, depending on the amount of memory available, and the type of peripheral devices attached.

Cassette tapes are commonly used as low-cost file-storage devices, but suffer from only being accessible in a sequential manner, so they can be very slow for some applications. Floppy discs provide random-access capability, and hence much better performance. Capacities range from about 60 000 characters for a single-density 5" (12.7 cm) diskette to about 1.2 million characters for a quad-density 8″ (20.32 cm) diskette. The discs themselves are very cheap, say, a couple of pounds each, but the controller and drive units cost several hundred pounds at present. Larger-capacity discs are coming on the market as the technology develops but it is likely that the backing store for future terminals will be solid-state devices, probably bubble memories, which will give better performance and higher reliability.

Other peripherals being developed include printers of varying speed and quality, and a very interesting card reader capable of reading the magnetic strip on the back of a credit card.

An alternative approach would be to take a computer and add to it videotex communications. This is already happening with hobbyist computers, giving the equivalent of a mid-range intelligent terminal. The top end of the range is a business computer. Then, if we consider a multi-user configuration, this could be a private viewdata system in its own right. Thus we would have a growth path from a simple terminal through to a stand-alone system.

Engineering flexibility

Before opening up the wide world of applications for 'intelligent' terminals, however, a number of engineering problems should be noted. These must be overcome for passive or 'dumb' terminals and intelligent machines alike if the ideal of full flexibility and mass-production economics is to be attained.

The standard videotex 'chip sets' referred to earlier deal with the interface between the communications equipment associated with the terminal and the basic CRT (cathode ray tube) and electronics drive.

The digital character stream received from the videotex service (or teletext) is dumped into a buffer (almost certainly a standard microprocessor with a small amount of local storage) and then read out again to a video generator which produces the right signals to drive the normal analogue electronics which, in turn, control the colour and positioning of the light spot on the CRT screen.

Within the chip set, a set of standard characters are stored in the form of coded instructions to move the spot, e.g., to draw an 'A' on the screen.

The complete image on the screen is constructed from a limited number of scan lines, and to ensure legibility a reasonable number of these scans must be allocated to each row of characters in the display format. Prestel is divided into 24 lines of 40 characters each. Many languages require accents and this means more scan lines must be set aside. In addition special national characters are needed, e.g., in German, plus all the special sets of symbols for mathematics, etc. All this presents a severe problem for International Standards committees in agreeing coding conventions. The contribution that the chip manufacturers can make is to support many different character sets locally, thus avoiding the need for special hardware for each country and/or application and keeping transmission times to a minimum. The host computer system would only need to send an occasional signal to specify the character set to be used, instead of transmitting long-winded codes to specify one character out of an enormous range for each position on the screen.

More flexibility can also be given by building in reprogrammable stores (EPROMS). There are read-only memory chips, used for character-set storage, that can be loaded or reprogrammed at local-supplier level.

In Prestel, spaces appear on the screen where the colour, say, changes in a line. Hidden in this space is a colour-control character embedded in the data stream coming down from the computer. If more information were allowed, e.g., two bytes instead of one for each character position, then the attributes, such as colour, flashing status, etc., of each character as well as its identity could be set without leaving spaces on the screen.

The penalty is more transmission time and scope for transmission errors. However, hardware trends will minimize these drawbacks before long and also lead to the ability to hold high-resolution symbols locally, i.e., to facsimile standards.

On the same theme, it is frustrating not to be able to use the full resolution of the TV screen in order to incorporate pictures into videotex frames. High-speed communications (see below) and better control organization will make this technically possible soon and indeed it is already specified for one national videotex network.

Dual-purpose VDU–videotex terminals

The cost of colour TV monitors is decreasing quickly and high-quality 14–16″ (35.56–40.64 cm) colour sets are available for business videotex applications, i.e., desk-top terminals. Two improvements are vital. Firstly, flicker must be eliminated. The latest designs achieve this. Secondly, for the business user who wants to access his company database normally via a VDU, and to take advantage of videotex information services on a colour TV, a way must be found of integrating the two functions in one desk-top terminal.

The latest developments argue that for a relatively small increase in cost and with careful design it will be possible to switch the terminal between different formats by transmitted program command or from the local keyboard. Such terminals are now on the drawing board. Only one final piece of hardware is then needed to achieve a really flexible terminal—a local programmable micro-computer.

In practice, 'intelligent videotex terminals' have already been simulated for IP editing purposes (see below) and telesoftware experiments simply by connecting a micro-computer externally to a videotex terminal. This gives enormous scope for developing videotex services.

There are no engineering reasons why the micro should not be put 'on-board' inside the TV along with the other 'chips' as soon as the manufacturers are convinced that the market exists.

Telesoftware

When we have a reprogrammable device, i.e., a computer attached to a communications network, such as a videotex service, it is possible for the network to be used to distribute computer programs as well as information. This

59

is the telesoftware technique which provides a major extension to viewdata systems in adding computational facilities by the use of intelligent terminals. At first sight it would seem simpler to provide this capability at the central computer. However, with the increasing capability and decreasing cost of microprocessors, such an approach to the time-sharing system appears less valid. With a significant number of simultaneous users, the processing power of a more powerful and expensive computer is needed, which contradicts the viewdata philosophy. Telesoftware leaves the central machine to concentrate on communications (for which it was designed). The special processing takes place in the terminals which are only dealing with a single user at a time and can therefore have adequate processing capacity.

Applications of telesoftware

The uses of telesoftware may be considered as belonging to two categories: programs which are used when the terminal is on-line to the viewdata system, and programs used off-line, independently of the viewdata system. The first category, where programs operate on-line, can be considered as a means of enhancing the capabilities of the viewdata system as a whole. Examples of this include:

* *Improved methods of access.* The tree search method which is the only standard method on viewdata has the important advantage of being very easy to use and completely general. It is, however, rather clumsy under some circumstances; for example, when applying multiple-choice criteria. While technically possible, the incorporation of such selection mechanisms into the central system would create another special database system with access methods to be learnt, and would destroy the simplicity of operation. By making use of an intelligent terminal, it is possible to have special access methods tailored for individual databases, while retaining the standard method for general use.

* *Processing of data.* The value of information on a database is greater if the user has the ability to process it. A simple example would be the ability to calculate the value of a portfolio of shares from their latest set of values. However, this facility is of most use in a closed user group when the amount of data is fairly small—for example, a head office sending latest price information and delivery schedules to subsidiary companies.

* *Validation of response frames.* Prestel does not carry out any validation of data entered on response frames (again to keep down the central processing overhead), so errors are not detected until the information provider retrieves them. With an intelligent terminal it is possible to ensure that only valid data is entered, for example, by subsidiary companies placing orders on their head office.

In the second category of the use of telesoftware, where the programs themselves run off-line, viewdata is being used purely as a distribution medium, but with the advantage of a very wide user base, and low cost communications. Examples include:

* *Accounting programs for small businesses.* There are many very small

businesses, such as corner shops, which could make use of accounting programs if they could be 'rented' at a low cost. This market for viewdata telesoftware could become very profitable.

* *Training and education.* The use of colour and graphics capabilities on an intelligent terminal can provide a very attractive medium for computer-assisted learning. Off-line usage means that the student can go through the course material at his own speed, but at the same time the communications aspects of viewdata can be used to ensure that he always gets the most up-to-date training material, and if required, can be used to send back results of tests to a tutor.

* *End-user computing.* The colour and graphics capabilities also make the intelligent viewdata terminal a very useful tool for the end user in a business organization wanting to do his own computing. For example, he might use it to present financial results in graphical format, to study trends, etc., and the combination of capabilities of both a small computer and a viewdata display terminal results in a very attractive device.

Standards for telesoftware

Videotex systems in general are still under development, with the Canadian Telidon system, for example, having significantly more advanced facilities. In the same way, we can expect facilities for intelligent terminals that are not present on standard viewdata terminals. For example, the standard monochrome display has 80 characters, and 80-character colour displays are already becoming available for the less price-sensitive market.

With telesoftware we must distinguish between the standards used by the transmission medium, and those used by the device executing the program. Prestel, for example, will be used to transmit programs which can optionally use non-Prestel-like facilities. Similarly, alternative formats are possible for transmission. The seven data bits plus parity is not optimal for telesoftware which needs to transmit eight-bit characters, but conversion to different transmission formats is relatively straightforward.

A much more serious problem arises from the range of different intelligent terminals and the different programming languages they employ. If a given telesoftware program can only run on a single type of terminal, the facility is only likely to be useful in closed user groups. To overcome this, high-level languages are being developed that can give identical results when run on different machines. UCSD Pascal, and CAP-CPP's MicroCobol are examples of this; the latter is already being used for experimental telesoftware on Prestel.

A further problem area concerns copyright and 'anti-piracy' protection. This is a general problem for all information on a viewdata system, but it is possible with telesoftware to incorporate such protection in the form of an expiry date for the programs as they are brought down. Hence there is the possibility to 'rent' software at a relatively low usage charge, which will ultimately lead to a mass market.

61

Editing

In the beginning, the pioneering Prestel IPs were provided with an on-line editing capability. Its functions and facilities were minimal. The software supporting it was bug-ridden and unselective, so that all-too-frequent line errors played havoc with the appearance of lovingly and laboriously created frames. Editing was a tedious, exasperating and often soul-destroying occupation. But the Post Office was mindful of the need to encourage and support its providers of information—so editing time was free, telephone charges were at local-call rates, and the hand-crafted editing keyboards were available at a nominal rent.

Even though the keyboards only supported the original Mark I set facilities, and had no background colour, double height, or separated graphics, and even though they were like gold dust to come by, yet they existed and were cheap.

Editing in those days was not a question of database creation. Information had to be classified, divided, and subdivided into displayable units built up into coherent structures mapped on to the simple Prestel trees. Structures became frozen because there was nothing available to help with rebuilding and recording.

At that time investigations started into the development of off-line intelligent editing systems both by set manufacturers and writers of software. The hardware was assembled by connecting a micro-computer between the Prestel modem and the TV set. This meant using either a Mark I set with external modem and no new facilities, a specially adapted set with a V24 interface, or a second modem back-to-back with the integral TV modem and a third modem on the Prestel line. The micro also needed to be approved for connection to the public switched telephone network. This made prototypes bulky, unattractive, and costly.

Studies were carried out into the functions required by IPs of an intelligent editor, and the result of these indicated a desire for considerable help with on-line editing and a means of editing off-line, thereby avoiding line errors and giving an enormously increased measure of security. Unfortunately for intelligent editing terminals, the Post Office policy on charging had already distorted the market. IPs by this time had differing requirements and were not prepared to fund the development of the comprehensive package which would fulfil their collective needs. With the limited size of the IP market, and its demand for cheapness, set manufacturers and software houses judged that the investment was too great a risk, and editing progress entered the doldrums.

The Prestel IP market is still of limited size and not on its own sufficient to support the development of very sophisticated systems. For this reason, IPs who wish to increase the diversity and quality of their database and its presentation are moving tentatively towards bespoke systems, while others, especially those with smaller or less volatile databases, are content to wait for the eventual appearance of a limited-function intelligent set.

So far, the emphasis has been on Prestel, but there is the business world to consider, with its as-yet embryonic private viewdata systems. This market is potentially enormous; private systems could outnumber Prestel centres in next to no time. The expectation of quality and the investment capacity are far greater than in the IP market place, because management sees viewdata as a way of cutting existing costs and/or improving the efficiency of their information dissemination and the timeliness and quality of the information. IPs, on the other hand, have no potential cost savings in using Prestel, only real costs.

But a viewdata system for the business market will ultimately stand or fall on the convenience, speed, and accuracy of getting information into it. The normal Prestel standards will not be acceptable to the business user, who is accustomed to sophisticated, user-friendly dialogues, formatted displays and data entry forms, field and screen validation, and similar techniques supported by business computer systems.

Where the IP is full of zeal and will press on against all odds, the data-entry clerk will, once the novelty has worn off, throw up his hands in disgust and take himself elsewhere. The producers of private viewdata packages have appreciated the need for easier editing, and have all provided extra and improved facilities. However, the facilities which can be provided by the central system are of necessity limited if the cost of hardware and software is to be kept to a realistic figure. This need to keep the processing load on the enquiry machine to a minimum once again throws the editing requirement back out to the terminals. Viewdata frame preparation is an ideal candidate for distributed processing, especially as the large range of alphabetic and graphic characters and colours allow numerous combinations and formats in the presentation of material. It is easy to see that different types of editing terminal—or multi-function terminals— will be needed for comprehensive private viewdata systems: terminals for designing the basic frame formats and graphics components to be associated with a given database; terminals for colour word processing; and terminals for formatted data entry with field and screen validation.

To gather the information for an effective private viewdata system, one may imagine every typewriter replaced by an editing terminal. On such editing terminals, whether they are intelligent terminals or terminals connected to a more intelligent editor provided by the central system, the keyboard is an important feature. Most of the current keyboards provide only the characters which can make up the viewdata frame together with the characters or character sequences which have special significance to the Prestel editor. Many of the characters which are frequently used by computer systems to indicate control functions, delimitors, and terminators cannot be generated by these keyboards. Using such a keyboard for more intelligent editing therefore entails a longer, clumsier dialogue to indicate what functions are required. Where display characters are concerned, many keyboards have disadvantages: for example, the absence of a 'shift lock' key and the wasteful duplication of colour keys for both graphics and alpha colours. The graphics keypad, elegant in concept and simple to use as it is, nevertheless is painfully slow in practice. The provision of function keys, backed up by powerful editing functions in the terminals or system, could dramatically ease the task of building pictures for viewdata.

Screens for business viewdata systems will also need improvement, especially on the editing side. It is far less important for the business user that he should be able to receive broadcast television than that the screen should be flicker-free. It is

63

far less important that he should be able to see the screen from an armchair twenty feet away than that the resolution should be comfortable and that a single screen should present enough information at one time.

So what do these speculated requirements imply? More keys on the keyboard? More characters on the screen? Certainly, such terminals will be needed. Keyboards and keypads with more functions, 80/40 character software-switchable displays—these will have to come about if the viewdata concept entire is really to penetrate the business world, and not merely be mapped on to colour VDU's.

As well as the changes in terminals, changes will be needed in computer software to support business viewdata systems. Computer language support for screens, colour, graphics design, and viewdata databases must also make its appearance. Mainframe links for the transmission of existing data and associated updates with automatic incorporation into the existing database will tend to become the mode of operation for businesses with significant data-processing operations.

Where, then, are editing terminals going? For IPs at least, progress can only be fuelled by Prestel. The current simple interactive keyboard will eventually give way to limited-function intelligent editing terminals, with or without local off-line storage capability. For a large operation, movement can only be towards off-line systems. Such systems will typically be mini-based, and may support dumb editing terminals, intelligent terminals working in on-line mode, and links to mainframe and other sources of basic information.

In the business world, the possibilities are far greater. Here terminal communications technology is already well understood. Security protocols, word processing, exchangeable character sets, and colour graphics are all commonplace. In the US, colour VDU's are now available at very little more than the cost of similar black-and-white terminals here. Apart from broadcast TV reception, the VDU manufacturers are better placed to make a speedy move for the business editing terminal market. Will they beat the set manufacturers to it?

Videotex and communications
Network growth and videotex
Videotex is no more than one among many new business communication techniques. The expansion of these ideas is linked to the very strong growth in recent years of communication services. It has been said that Parkinson's law applies to modern communications—the traffic expands to fill the channel capacity available. Certainly, intensive development of data transport systems and switching equipment will continue. However, increasing attention must be paid to the problem of giving the user proper control and access to all the data becoming available.

Videotex is one answer. Although unsophisticated in comparison with other information retrieval techniques, it represents an attractive man–machine interface which is likely to prove very popular. Success will, however, depend on full exploitation of modern communications technology. In these early days of videotex development, 'communications' is regarded simply as a subsystem within a single private or public videotex system. In the long-term there will be links between different videotex systems and eventually between information retrieval and computer services of all descriptions on an international basis.

Developments in optical fibres, satellite relay stations, and wide-band switching systems will eventually put almost unlimited bandwidth, i.e., information-carrying capacity, within reach of everybody and communication costs will be small compared with the cost of the data.

Unfortunately, this utopia is going to take some years to materialize and in the meantime communication charges will form a noticeable part of any videotex user's or system operator's bill. Most systems will depend initially on the existing installed network of copper wires and at least to some extent on the relatively low-quality switching devices connecting them together at the PTT exchanges.

Communication technology at local level

The terminal components mentioned so far are the standard CRT electronics; the videotex chip set, the local micro-computer, and the special electronics for switching between VDU/videotex operations. An alphanumeric keyboard rather than a numbers-only keypad is required to exploit the local computer and is likely to be favoured even for 'dumb' terminals as the range of applications grows.

This leaves the communication hardware, which means the electronics, including usually a 'modem', which terminates the telephone line and supports the exchange of control and timing signals establishing and maintaining the dialogue between the terminal and the host computer.

The signal speeds of Prestel are 1200/75 baud, which implies about 8 seconds to fill a screen. Modems offering 9.6k baud transmission over ordinary unequalized, switched telephone lines have recently become available and data rates of 64k bits per second for each local terminal are planned for future services such as Telidon.

In conventional data processing and for bulk data exchange between videotex centres, it is already common practice to provide automatic transmission-error detection and correction. With the current state of most public lines it is important that such error protection is brought down to individual user level.

The low cost of microprocessors and corresponding incorporation of them into 'intelligent' devices from modems to terminals will make it a straightforward technical task to support the necessary error-handling software locally. Business users will then be able to obtain high-performance services free from data corruption without the need for expensive leased lines.

Exploitation of the local terminal computer, telesoftware applications, and the advent of more powerful components such as megabyte bubble memories and high-performance peripherals such as video discs will make faster telecommunications essential rather than desirable.

The user will see better screen response, the network and the host computer

65

input/output ports will be used more efficiently, and this should indirectly result in lower costs all round.

Within both the videotex service computer(s) and the local terminals for users and IPs, there is a need for designers to balance exploitation of the basic speed of the hardware with flexibility. Flexibility is needed to take advantage of new hardware developments without re-engineering the whole system and also to handle the demands from users for new applications. This has led to the decoupling of 'processes' and 'services' within the software.

A program may carry out an application function such as modifying a text record. This is a 'process' and to support it 'services', e.g., a transport service code module, fetch and carry the data to be operated on and interface to specific hardware devices.

Thus in principle a designer can optimize his system and adapt it to use the fastest modems, etc., that the communication networks can support. A videotex system and individual terminal must both be able to interface to a range of other systems to meet user demands for an all-purpose desk-top to computer support. This introduces the problem of standards again.

System-to-system interfaces

The attractions of universal standards are clear. Already in the UK, user terminals can access both private videotex services and Prestel, to everybody's benefit, because there is *de facto* agreement that Prestel standards should be available on all systems even if only as a sub set of a more sophisticated private service.

Compatibility problems can be split into two categories: (a) the man–machine dialogue and high-level structure of the information database; (b) the technical interface between different communication networks.

The very serious problems arising at an international level in category (a) are treated elsewhere. Designers, by building 'soft' systems, can overcome certain incompatibilities by loading alternative software into the relevant subsystems and translating messages from one system format to another. There are likely to be severe penalties in cost and performance and where the system differences are more than skin-deep, e.g., the frame resolutions are different, prospects for a commercially viable interface are poor.

At the technical device/network level, (b), the videotex compatibility problem merges with the general problem of establishing universal communication protocol standards.

There is likely to be agreement on a seven-level hierarchy. The lowest levels will set the conventions for the electrical interface and moving up will define the 'handshake' protocols for establishing communication paths between different devices and systems. With the proliferation of new digital services, the need to agree on common standards has become urgent. CCITT recommendations for some interfaces, e.g., V24, have already become general standards but solid agreement at higher levels, e.g., V25 for packet switching, is not so easy to attain.

Packet switching offers distance-independent tariffs to the videotex user whose messages, reduced to digital form, are no different in kind from other digital traffic. Many PTTs, for example, will be offering such services in the near future.

Since agreement on standards will take time, designers must tackle interface problems on an individual basis. Overcoming such problems should ideally be transparent to the user. This has been done in certain cases already. For example, 'Seaview' is a videotex system, Prestel-compatible, which happens to use radiotelephony for linking the terminals to the database. Less trivial is the Fintel system, where information may be exchanged between established on-line retrieval services via Euronet, Telenet, etc., a local mainframe information service bureau and Prestel-compatible videotex systems. This involves a central small computer to handle code and file conversions, etc., and to support multi-purpose workstations. Such a station may be used to edit material for input to the videotex database or other on-line systems, for interrogating any services, and to maintain local security files, housekeeping logs, etc. In due course, much of this could be handled in the local terminal without an external support processor, and programs for specific interface tasks might be maintained by telesoftware techniques.

Another approach which can be straightforward is 'emulation' where a common commercial interface at low-level (CCITT 1–3) standard, e.g., IBM 2780, exists. The terminal and host system both have this mode of operation available and a 'foreign' terminal can hook itself up as if it were a terminal of the type normally connected to the host network.

It will be sensible in the future to use telesoftware to load the appropriate interface software from the 'home' videotex service or a third-party supplier, that is, to tackle each interconnection problem as it arises.

Office systems
Information management
The function of the office is exclusively concerned with information. Office workers, whether they be typists or administrators, spend most of their time preparing, communicating, and retrieving data. Collectively they form a massive overhead cost for the average business. Company directors are, as a result, becoming more conscious of the cost of information management at the same time as they are individually coming to appreciate the value of timely and accurate information in a competitive world. This climate has led to the growth of the office automation market which will be a vital application aea for videotex services and techniques.

The designs for some of the first complete office systems include videotex ports, and there is no doubt that most of the individual terminals in the office of the future will support, among others, a videotex mode of operation. The videotex option may be used for access to a videotex service, e.g., Prestel; almost certainly for messages and electronic mail; and finally as a file-management technique enabling a non-programmer to make effective use of a computer to maintain his own database on behalf of his work-section or department. It is worth looking briefly at how this is likely to be implemented.

Terminals

Analysis shows that office workers may be divided into three groups: typists; clerical and administrative support; executive and professional. Computer support shows promising cost-benefits for each category but most of all for the executives, the fastest growing and most expensive sector of all.

Typing terminals

'Word processors' are now common. Good communication facilities are now being added to the basic VDU–keyboard–floppy disc–printer unit, and those devices are likely to provide effective support without any requirement for specialist techniques such as videotex. (As an aside it is interesting to note how the established screen-based word-processing techniques for text handling are being adapted for videotex frame-editing systems.)

Clerical support

Hardware similar to the word processor is likely to be suitable here, but it will be more versatile with local intelligence from the beginning. Few 'clerical' terminals have been designed so far because of the impracticality of asking non-experts to handle the average computer database management package, and clerical work is mostly about filing and retrieving information. The hardware for raw data storage and swift access is rapidly becoming technically and commercially viable at the departmental level, but the software and indexing packages available are too complex for the non-programmer.

This is where the videotex approach is likely to be welcomed as a technique for indexing and structuring data in a very simple way. Although unsophisticated, the simple search mechanism implicit in videotex tree structures can enable the unskilled part-time operator to use computers to maintain indexes and manage generally a sizeable local database—a large step towards the paperless office. Local intelligence can be used to improve the retrieval performance (keywords, etc.) and for validating data which the clerk may send to a remote system via videotex response frames.

Executive terminal

Colour TV screens are excellent vehicles for presenting complex information. Executives spend much of their time retrieving and assimilating data. The combination of new low-cost colour screens with public and private information services, particularly videotex developments, is leading to a new market: executive communication terminals.

In financial centres it is already common to see professional staff use video terminals for on-line access to market data. In London there are several examples of multi-service terminal systems under trial and Prestel is an option on all of them. The protocol is simple to understand and remember, which leads to a demand for an equally convenient access to be provided to the central computer files of the business.

This requirement is matched by the clerical staff, who besides maintaining local files need to access the central system. This means that a route, direct or indirect, must be provided to at least segments of the company database via standard videotex commands. It might not be necessary to dress up the data in full colour videotex frames, and here the switchable VDU/videotex terminals will be invaluable.

Given the rapid fall in hardware costs, it is difficult to see executives ignoring the advantages of an on-board micro-computer. The local intelligence will also support software for calculations, data extraction from videotex frames, histogram and graph generation, and personal services such as diary keeping, message handling, etc. (the 'electronic in-tray').

The office system

One aspect of office automation is already evident: automation will not be allowed to upset the traditional organization of jobs in the office for many years. There is a consensus emerging among designers about how the various terminals will link up into integrated office systems. Briefly, this will depend on a ring main either physically constructed as such or implemented within a powerful message-switching black-box similar to a sophisticated local telephone exchange.

Small systems may involve no more than half a dozen communicating word processors, a shared printer/copier, and a port to the local data-processing centre. A port to a videotex service may be common soon after as the system evolves. It depends obviously on the nature of the client business exactly when the need will arise. In larger systems, a local semi-public videotex service may be supported, that is, software running on a mini-computer attached to the ring as a shared resource for all terminals. Other small computers on the ring would provide processing power for terminals without suitable local capacity, and would handle telex interfaces and electronic mail services to other office systems and so on.

Finally, the 'back-end' database processor handling bulk file storage and retrieval for the network is likely to support a videotex protocol, as an option if not the only means of access. The advantages of adopting the same protocol or a subset are apparent. The user, whether a clerk or senior executive, can store and retrieve information using private, departmental, company, commercial, and/or public database through one easy-to-operate man–machine interface.

The key to the development is not technical but psychological—the familiarity of a colour TV and the telephone and the widespread acceptance of a method of indexing and structuring data. Prestel and the later videotex systems are likely to be the *de facto* standard, which puts pressure on the terminal and system suppliers to make their products flexible and compatible with these standards. In return, the office market will provide a strong stimulus to products of all kinds.

The future

Public authorities are planning to use videotex services in the domestic environment for everything from energy management, meter reading, alarm systems, to multi-channel broadcasting and bespoke education. Public services, e.g., French directory enquiries, may be available only via videotex, and with network and terminals established, private companies from publishers to mail-order houses and banks will start to exploit the system. However, with the exception of some notable pilot trials, all this will take place some years in the future, due largely to administrative and financial factors, in particular the non-availability of a suitable low-cost wide-band communication link into the average house.

On the business side, independent systems for data collection and private information services are with us already. The office badly needs an alternative to the mail and telex systems for store and forward correspondence and for local file management. Many executives dependent on access to vital fast-moving data, e.g., market information, would welcome convenient desk-top access to the services of information brokers, etc.

Technically, it has been indicated that a really flexible terminal can be designed and, given hardware trends, should be economically viable very soon. Sophisticated editing systems are becoming available making the task of database creation and support reasonably efficient and the PTTs are now installing powerful digital communication networks which could carry the national and international videotex traffic. So what are the problems?

Two are technical. Firstly, standards. Without agreement, designers will not know how to ensure inter-system compatibility and the all-important evolution to high-resolution systems making better use than Prestel of the intrinsic resolution of the colour CRT. Secondly, very large-scale system design. Relatively few systems of the scale envisaged by the BPO and other PTTs have been built. Much work has still to be done to ensure adequate security and operational performance on systems supporting hundreds of thousands of on-line terminals.

That said, there is no doubt at all that these technical problems will be overcome, as will others still to arise. The world is developing a great thirst for information and there is a substantial drive coming from high-technology manufacturers who see computers applied to text, as opposed to numbers, as the source of future expansion.

A factor restraining growth in the new 'information management' systems has been the historical tendency of system designers to make people fit machines. Videotex is a welcome break from tradition and is likely to succeed where more conventional approaches have failed. The business world is starting to take it up now, new intelligent terminals will provide further encouragement, and the domestic market seems likely to follow 3–10 years behind, depending on the country.

4. International technical standards

Richard Clark

Introduction

Although the majority of national viewdata services look similar, there are basic differences in opinion between implementers as to what constitutes a viewdata system. The Prestel team in the BPO took a marketing decision to freeze development in the mid-1970s and to introduce the basic service as it existed at that time. Other countries delayed the implementation of their viewdata services for four or five years, and consequently were able to offer sophisticated services making use of more advanced technology, but without the benefit of several years' field experience, or indeed proof of their technical abilities.

There are many reasons why these differences and delays occur. They relate to the perceived importance of viewdata by those responsible for development or service introduction, the comparison with other similar services, and the costs of implementation. Without doubt, one of the fundamental parameters of a new telecommunications service is its attractiveness to users, and this, together with political factors, affects the potential for export.

The rewards offered to a country which can develop a new, and patentable service are immense. This in turn leads to very significant political pressure being brought to bear on other countries to adopt the service. Dr Rhonda Crane (1978), in an excellent review of the issues affecting the adoption of a standard for the encoding of colour television signals (NTSC *v.* SECAM *v.* PAL), points out that for the *loser* in this particular debate (if indeed a loser exists), the French SECAM system, the export market in 1975 for French colour television sets alone was valued at nearly 28 million dollars. If the sale of technical assistance and support, specialist studio and transmission equipment, and programme material is added to this figure, the export potential for SECAM was certainly in excess of 100 million dollars per annum at this time.

In some ways it is unfortunate that international technical standards should be dragged into the struggle for service definition. Many of the purposes of international standardization, such as reduction of costs to the user, are undermined by the political emphasis placed on success by one country in presenting its case. Nevertheless, adoption by a telecommunications service can have lasting economic benefit to its inventors, and any observer of the international standards scene must have this fact in mind.

This explains to a large extent the importance that technical standardization has in the adoption of international viewdata standards. This chapter concentrates on examining the need for it in relation to:

● the benefits of standardization for viewdata services

- who is effective in determining technical standards for viewdata
- the needs of viewdata in standardization
- possibilities for the future.

The need for standardization

There have been many debates on the issue of whether standardization is necessary, or whether it should be left to a competitive market place to establish the most successful commercial offering of a service. While the latter approach can have merit in producing cost-effective solutions which are technologically advanced, it is also expensive in overall development costs and can lead to market domination by individual suppliers. It places a great deal of emphasis on protection by patent, by secrecy, and by technological advancement and change. If viewdata systems are to be regarded as universal tools for information dissemination, utilized by a wide variety of suppliers and users of information, then these factors would tend to detract from the usefulness of the service. If, on the other hand, viewdata services are to be designed for specific user groups, and are viewed as a group of separate and competitive offerings, then individual unstandardized service development may be attractive both to user and supplier. This difference is most apparent for the business viewdata user, where there is competition between the standardized offering of the PTT or common carrier, and a variety of private viewdata systems developed to appeal to specific user groups.

There is no real need for conflict between the protagonists of standardization and those who oppose it. The issue is rather one concerning how compulsory should standardization be, and how should compliance with a standard be examined. Many standards are little more than a ratification of a *de facto* standard created by widespread acceptance of a commercial offering. Most telecommunication services involve standardization at some level in their design. It is extremely unlikely that the most vehement opponent of standardizing a viewdata service would choose to ignore basic standards on safety, on ergonomic design, or on interfacing the different components of his service. He would, however, feel that any attempt to categorize a viewdata service as a collection of standards was fundamentally wrong.

Why are these basic standards chosen? There are a number of reasons for adopting technical standards:

- to lead to economies of scale in the production of equipment
- to improve the ease by which different components of a system may be interchanged
- to help define a system as a number of modules which may usefully be interconnected
- to ease problems of introduction, training, and education of the people involved with a system
- to allow freer international trade
- to simplify use by a reduction in complexity through fewer variants
- to help in system specification for procurement or legislative purposes

- to maintain the quality of the component parts of the system
- to allow future enhancement of the system without excessive cost penalties.

It is doubtful whether an implementer could point to all these benefits for any one service, but individual examples of the benefits, or counter-examples of disbenefits occurring through non-standardization are common-place.

A most important factor is the question of economic benefit. A number of studies have been carried out which examine the cost/benefit analysis of standardization. In France, IRIA, a national research body, reported that over a ten-year period the savings of using a standard programming language would amount to perhaps 6 million dollars, as against a cost of production of the standard of around $1\frac{1}{2}$ million dollars. The use of standard interfaces allowing so-called plug-compatible manufacturers to interface to large computer systems has been estimated to have saved the US Government over 120 million dollars. Calculations by the NCC (1977) in England indicate that saving of a mere 1 per cent of the operational costs of computer systems would recover the typical investment in an international standard (about 4 million dollars) in one year.

Standards bodies

There are a large number of organizations, both national and international, which have significant influence on standardization activities. For viewdata services, the major bodies involved are:

- international standards bodies—CCITT, ISO, and less directly CCIR
- international 'affinity' groups—CEPT, EBU
- national standards bodies.

International standards bodies

CCITT is undoubtedly the most influential body so far as viewdata is concerned. The initials stand for the International Telegraph and Telephone Consultative Committee (actually referring to the French translation of the name, and showing the emphasis on both official languages in the CCITT and the International Standards Organization (ISO)).

CCITT, together with CCIR (International Radio Consultative Committee) who deal with standards for radio and television, are both part of the International Telecommunications Union (ITU), which is a direct agency of the United Nations. Its major objective is to produce a variety of recommendations dealing with administrative, tariff, and technical details concerning international telecommunication. Membership includes not only the PTTs (postal and telecommunication authorities) of interested countries, but a range of government departments, scientific organizations, other international standards bodies, or operating agencies. Service users, however, are not directly represented. CCITT operates in a four-year cycle, each cycle being terminated by a plenary assembly which can adopt or reject recommendations placed before it by its study groups, and can also generate further questions for study in its next plenary period.

Two study groups are currently the main parties interested in viewdata

73

standards. (The CCITT and other international bodies have adopted the generic term videotex rather than viewdata to avoid confusion with the original UK development.) These are

- SG I—whose function is to define the service and the facilities it offers.
- SG VIII—which is interested in technical specification.

These study groups in turn are further subdivided into working parties, and rapporteurs groups, in order to speed the work flow. CCITT has been involved in viewdata standards since May 1978. Since the 1976 Plenary failed to enter a question on viewdata, perhaps having underestimated the interest in and importance of the subject, a special *ad hoc* meeting had to be convened to place viewdata on the agenda for discussion. This took place primarily on the instigation of the CCIR, who were attempting to standardize broadcast teletext, and felt that they needed guidance on the integration of teletext and viewdata. Since that time, working groups have met at approximately six-month intervals—with rapporteurs groups meeting even more frequently—and the number of contributions on viewdata has escalated.

The primary interest of the CCITT was originally to ensure that international telecommunications could be ensured with a minimum of difficulty. Gradually it has put increasing emphasis on defining those system components which make up the concept of a service—telephony, telex, or viewdata, for example. This has increasingly brought it into territory already occupied by the ISO, which is primarily concerned with a service as seen by a user and is made up of the many national standards bodies. ISO sponsors standards in a very wide variety of areas—over 150 technical committees already exist, each divided into numerous sub-committees and working groups. For viewdata, ISO has an interest in representation and layout of information, and in interfaces to peripheral devices. Most coding for characters is based on the ISO standard ISO 646, and the technical committee responsible for this standard, TC 97, is active in defining other viewdata standards.

International affinity groups

A number of groups exist which meet on an international basis, but restrict membership to a particular type of organization or geographic region. These tend to form strong pressure groups in international standards bodies, and are of considerable influence in dictating future standards. Most significant for viewdata currently are:

- CEPT—a group of the majority of European PTTs
- EBU—the European Broadcasting Union.

Because of the need for close technical integration between teletext and viewdata, it is hardly surprising that the EBU has been involved in the definition of the basic facilities offered by a terminal. As potential service implementers, however, the CEPT has probably been the most influential body involved in international standards for viewdata. The tendency of the European PTTs to discuss and consolidate their viewpoint in private, before presenting a common front to CCITT, has often led to dissent at the CCITT, where the manufacturers, or the

USA as a major implementer, often seem to feel aggrieved at such behaviour. Although there has been much historical conflict between the UK and French administrations over standards for viewdata, once again, at the time of writing, the European PTTs seem to be reaching agreement on basic standards for viewdata.

National bodies

In each country involved in viewdata there are a number of bodies who represent the interests of various participants in the service, and serve to influence the national vote at CCITT and other international bodies. Thus in the UK, the British Radio Equipment Manufacturers Association (BREMA) represents the interest of the television industry, the Association of Viewdata Information Providers (AVIP) the information provider, while the vote of the British Standards Institution at the ISO supposedly represents user interest.

In other countries similar organizations exist and influence the standards-making process. The major national body influential in this sphere is undoubtedly the American National Standards Institute (ANSI), which has dictated a number of standards in the information-processing industry. The US version of the ISO 646 code, known as ASCII, is far more widely known and used than any other national variant, for example.

As well as the organizations mentioned above, there are a large number of other bodies who have professed interest in viewdata. These include the European Computer Manufacturers Association (ECMA), the International Electro-technical Commission (IEC), and the International Press and Telecommunications Council (IPTC). In total, there are almost certainly more than one hundred national and international bodies currently discussing viewdata standards. The division of responsibility between these different groups has never been clarified, and this leads to many difficulties. Not least among these is the tendency of the major standards bodies to finalize a standard or recommendation which includes many options—basically whenever agreement has not been possible. The worth of such a standard is questionable, as testing for compliance with the standard is very difficult to carry out. Unfortunately, it may well be that standards for viewdata will turn out to have similar failings.

Technical standards for viewdata

The first question to be raised when considering viewdata standardization is always, 'what is viewdata'? The definition of the basic service has evolved from that of a limited subset of the original British Post Office viewdata software to include a variety of features which were not felt economically viable at that time. As each new feature is contemplated, it is placed as a subject for study on an appropriate international standards committee. Rather than contemplate the ever-growing list of 'essential' viewdata features as seen by any one implementer, it may be helpful to show some of the major areas in which standards are required for viewdata, and historically, how discussion has resolved the conflicts involved.

These main areas are:
- service definition
- display characteristics
- terminal characteristics
- user control procedures.

These have been selected as they have already been the subject of many international contributions on viewdata standards. Other issues are interworking with other services, call-charging procedures, network strategy, international access, and transactional service implementation.

Service definition

The definition of what constitutes a viewdata service is fundamental to any standardization activity, and yet it is one which is most difficult to answer. It has been easy to define it in broad terms—'simple for the user'; 'acceptable quality of reproduction'; 'inexpensive equipment'; 'economically viable infrastructure'; etc.—but without any quantification this hardly helps the process of standardization. Without user feedback, how is the standards expert to know what is simple—what quantities he can expect to sell with a certain price tag, or which features should be fundamental to all terminals?

This is the area in which most information is required early, and yet which requires full market research. Many countries have attempted to provide some of the answers through their experience with pilot trials, but these are expensive and can result in long delays. In the meantime, technical parameters must be laid down to allow the mass market for viewdata terminals to develop.

Perhaps the only solution is to adopt either broad, indefinite descriptions to define the service, or to adopt a rigid definition which will constrain implementers at every stage. This latter approach is unlikely ever to receive ratification internationally, not least because of differences in the national need for a viewdata service. The researcher who wants an answer to the question 'what is viewdata?' will have to look elsewhere than international standards bodies.

Display characteristics

A basic parameter of a viewdata service is what a screen of information looks like to a user. The viewdata screen is divided into a series of adjacent character positions, and two questions need to be answered:
- what can go into each character position?
- how many such positions are there on a screen, and how are they laid out?

The first of these questions is defined by the character coding chosen, and the second by the frame format. Since the original development of viewdata, character coding has been a basic difficulty. Its standardization will be discussed at length as an example of the problems which arise. The character code can be further divided into two areas:
- what repertoire of characters should be available?
- how should they be coded for transmission?

The original repertoire for Prestel was chosen to comply with that provided on conventional typewriters in the UK, and to maintain compatibility with broadcast teletext. It did not agree with the repertoire of characters used for data processing, and so was allocated a different set of codes for transmission. The original ISO standard for data-processing character codes, ISO 646, had itself been the object of considerable dissent, and the international reference version (IRV) shown in Fig. 4.1 has a number of character positions reserved for

				b_7	0	0	0	0	1	1	1	1
				b_6	0	0	1	1	0	0	1	1
				b_5	0	1	0	1	0	1	0	1
b_4	b_3	b_2	b_1	column / row	0	1	2	3	4	5	6	7
0	0	0	0	0	NUL	TC_7 (DLE)	SP	0	@	P	`	p
0	0	0	1	1	TC_1 (SOH)	DC_1	!	1	A	Q	a	q
0	0	1	0	2	TC_2 (STX)	DC_2	"	2	B	R	b	r
0	0	1	1	3	TC_3 (ETX)	DC_3	#	3	C	S	c	s
0	1	0	0	4	TC_4 (EOT)	DC_4	¤	4	D	T	d	t
0	1	0	1	5	TC_5 (ENQ)	TC_8 (NAK)	%	5	E	U	e	u
0	1	1	0	6	TC_6 (ACK)	TC_9 (SYN)	&	6	F	V	f	v
0	1	1	1	7	BEL	TC_{10} (ETB)	'	7	G	W	g	w
1	0	0	0	8	FE_0 (BS)	CAN	(	8	H	X	h	x
1	0	0	1	9	FE_1 (HT)	EM	)	9	I	Y	i	y
1	0	1	0	10	FE_2 (LF)	SUB	*	:	J	Z	j	z
1	0	1	1	11	FE_3 (VT)	ESC	+	;	K	[	k	{
1	1	0	0	12	FE_4 (FF)	IS_4 (FS)	,	<	L	\	¦	|
1	1	0	0	13	FE_5 (CR)	IS_3 (GS)	–	=	M	]	m	}
1	1	1	0	14	SO	IS_2 (RS)	.	>	N	^	n	–
1	1	1	1	15	SI	IS_1 (US)	/	?	O	_	o	DEL

Fig. 4.1 International reference version

'national' options. Thus ASCII, the US version of ISO 646, introduces a $ in place of the international currency symbol, ⤬, and a tilde, ～, in place of the overbar, ⁻. The UK version had a £ in place of the number symbol, ♯ , and, again, a $ instead of ⤬. The code table adopted for Prestel is shown in Fig. 4.2. It complies with ISO 646 in only using national options, but replaces the underline by ♯ in contravention of the standard. This was done for two reasons:

- the standard contained an anomaly which would not permit £ and ♯ to exist in any one variant of the code table, and both were felt useful for Prestel
- the underline symbol is probably better regarded as a control function, and should therefore exist outside the repertoire of 94 graphic characters. Without a backspace function, the Prestel terminal could not implement it.

When the code set was put forward for adoption internationally for viewdata services, it immediately created problems:

- it did not cater for accented letters
- it was biased towards UK requirements
- it did not conform to the CCITT interpretation of ISO 646, its recommendation V3, known as International Alphabet no. 5 (IA5).

Counter-proposals were put forward, based on extensions to IA5, which required the terminal to have the intelligence to combine two or more subsequent character codes to represent a single character. The UK then extended the Prestel code table in a similar manner, and the ISO committee were involved in an attempt to arbitrate between the method of coding. They in turn became interested in the problem of defining a universal for all forms of text communication. This work helped define the difference between the repertoire of characters which could be coded, and the method of coding them, which at least permitted codes to be exchanged between different countries providing a translation process was undergone.

In the meantime the Canadian delegates to CCITT had raised the issues involved in representing pictures on a viewdata terminal. The original UK concept had been to subdivide each character position into six smaller rectangles, each taking one of two colours. This approach, named alpha-mosaic by the Canadians, was added to by two new proposals—those for alpha-geometric and alpha-photographic encoding. Alpha-geometric coding drew pictures by using the terminal intelligence to draw lines or arcs between given points on the screen. This led to a higher quality of display, and had the added advantage that upgrading the terminal's resolution ability improved the image drawn, without requiring the entire database to be rewritten. Alpha-photographic encoding defined each character position as a series of dots, and allowed for higher-resolution pictures to be built up as a series of adjacent characters, although with consequent penalties in the transmission time of individual information frames.

While considering the implications of alpha-photographic coding, CCITT realized that many of the problems of rigidly defined sets of characters could be overcome by the concept of downloading. This method of coding sends a set of pictures, defined as dots, and occupying one character position each, to a terminal capable of receiving them, and from then on refers to each picture by a

Fig. 4.2 Viewdata transmission codes

Row	Bits b4 b3 b2 b1	Col 0 (000)	Col 1 (001)	Col 2 (010)	Col 3 (011)	Col 4 (100)	Col 5 (101)	Col 6 (110)	Col 7 (111)
0	0 0 0 0	NUL			0			—	p
1	0 0 0 1		Cursor On DC1	!	1	A Alpha" Red	P Graphics Red	a	q
2	0 0 1 0			"	2	B Alpha" Green	Q Graphics Green	b	r
3	0 0 1 1			£	3	C Alpha" Yellow	R Graphics Yellow	c	s
4	0 1 0 0		Cursor Off DC4	$	4	D Alpha" Blue	S Graphics Blue	d	t
5	0 1 0 1	ENQ		%	5	E Alpha" Magenta	T Graphics Magenta	e	u
6	0 1 1 0			&	6	F Alpha" Cyan	U Graphics Cyan	f	v
7	0 1 1 1			'	7	G Alpha" White	V Graphics White	g	w
8	1 0 0 0	Cursor → BS	CAN	(	8	H Flash	W Conceal Display	h	x
9	1 0 0 1	Cursor ↑ HT		)	9	I Steady	X Contig Graphics	i	y
10	1 0 1 0	Cursor ↓ LF		*	:	J	Y Separated Graphics	j	z
11	1 0 1 1	Cursor ← VT	ESC	+	;	K	Z	k	¼
12	1 1 0 0	Cursor Home & Clear FF		,	<	L Normal Height	↓ Black Background	l	‖
13	1 1 0 1	Cursor → CR		-	=	M Double Height	½ New Background	m	¾
14	1 1 1 0		Cursor Home RS	.	>	N	↑ Hold Graphics	n	÷·
15	1 1 1 1			/	?	O	← Release Graphics	o	■

code, rather than sending the picture each time. By this means any image could be built up on a screen or sent as an individual character. The standards discussion then changed from 'what should be the total repertoire of characters that a viewdata terminal could receive?' to 'what resolution should each picture element comprise, and how should it be coded?'.

Character coding has always been a major difficulty for viewdata and this rather abbreviated version of how it has developed shows not only some of the problems of standardization, but also the interdependence of standards which leads to the delay in their production. The other problem in defining the display—the format of the screen—has also led to many different proposals, but currently it seems likely that the difference between television standards in terms of the number of lines on each screen (525 and 625 being the most common) will lead to two standards for display—either 20 or 24 rows of 40 characters.

Terminal characteristics

The features of the terminal itself, and in particular those which must be available in the most basic rendering of a viewdata service, are another important discussion area. As far as a service user is concerned, the most important parameters are: whether the display is black and white, black, white and shades of grey, or full colour; what controls are available; and whether peripheral devices can be attached.

Again, service definition has not helped so far, and the standardization arguments show another drawback of international discussion—the emphasis on technical problems rather than user requirements. While the user is not generally concerned whether the modem was integral to his terminal, or what method the viewdata decoder used to connect to the television circuits, these are nevertheless the parameters which have come up for most discussion.

Many of the terminal characteristics will be defined by existing technical standards—layouts of keyboards, safety requirements, and network interfaces are examples. Development of new standards is not taken on lightly in areas where there is general agreement or an existing standard, and many such existing standards will be invoked in future viewdata services.

User control procedures

The way in which the user interacts with the telecommunication network to set up and clear down his call, and with the viewdata centre to request messages or other information is important, for it is by this means that viewdata is made simple for the untrained user. Here, however, much standardization is inappropriate, for a number of reasons:

- experience is the best guide to the methods that are easiest to use
- changing procedures only mean re-educating users, not replacing terminals
- it would be impossible, or at least inappropriate, to try and cater for all the requirements of current viewdata systems.

A basic set of control procedures, to connect a call, clear it, and request information, would be useful; and it is likely that procedures similar to those designed for Prestel could form this nucleus.

Viewdata standards and the future

From the examples given above, it can be deduced that the major effort given to viewdata standardization concentrates on defining a viewdata terminal. Partly this stems from an inability or unwillingness to define what constitutes a viewdata service, but in the main it is because this is where standardization offers the greatest benefit. In the longer term the major investment in viewdata will be in the terminal, and the current standardization activity is geared to ensuring as long a market life as possible for the next generation of devices.

The flexibility offered by down-loading of character sets, and the addition of full keyboards to the simple numeric pads can make each viewdata terminal a powerful interactive graphic display. There is plenty of room for enhancing this terminal by adding voice-message capability, longer line lengths, or local intelligence—perhaps even the ability to insert reduced still television-quality images into a viewdata frame. Incorporation of such facilities in a standard will require evidence of market demand, and so most of these features will only emerge after some entrepreneurial organization introduces them with such force that they become *de facto* standards.

The current activity on standardization has built on the experience of the UK to define a more advanced viewdata service. Political pressures have reduced the accuracy of the international standards by introducing many options into each, so that the major importance of each standard is the interpretation placed on it by a service implementer.

In the past, the standards which have been universally accepted, and widely implemented, have been those which are already *de facto* standards, and have been used successfully for several years. It does not appear that viewdata services will change this process. No one country is likely to be given the lead that full adoption of all its proposals would ensure. Standardization and politics are inextricably linked.

References

Crane, Rhonda J. (1978), 'Communications standards and the politics of protectionism', *Telecommunications Policy*.

National Computing Centre (1977), *Report on Standards in Computing*.

5. Viewdata and the television industry

Colin Tipping

To talk about television in the UK as a declining industry would probably seem to most lay observers to be more than slightly loony. Television growth has, in many ways, been a success story. British television programmes, from both the BBC and ITV companies, are admired around the world. Overseas sales of UK know-how are considerable. Technical standards are immeasurably better than even ten years ago. And despite occasional tiffs between TV managements and unions, the whole industry is committed to continued improvements in quality.

But when colour came to the UK in 1968, few predicted the massive lift-off in demand that was to occur between 1970 and 1974. British manufacturers were caught unawares (relatively) and supply companies found great difficulty in obtaining the sets they wanted in the UK. Inevitably imports increased. Equally inevitably, however, there followed a down-turn in demand and although colour growth continues at the expense of black and white (usually called 'mono' for 'monochrome' in the trade), it is pretty clear that something needs to be done to liven the industry up again. At present about 98 per cent of households have TV and about 70 per cent of these have colour. At least one-third of households have two TVs and there is a great deal of interest in the market for small, portable units. You can now put a TV in your pocket, and one that you can put on your wrist is not impossible in the next decade. But the fact is that the mainstream business for the UK TV industry is the set you have at home. The most popular size is a 22-in (56 cm) screen model.

It is said that every product has a life cycle which can be represented by a letter 'S'. This is because, if a graph is drawn of growth (upwards) and time (sideways), the graph looks like an elongated S. In the early period, growth is slow (the bottom of the S). In the 'boom' years, growth accelerates (the middle of the S). Inevitably growth declines as market saturation is reached (the top of the S). The TV industry is now moving well into the top of the domestic TV set growth curve. Once every household has a colour domestic set, manufacturing falls to 'replacement' level—that is, the only TV sets that are sold are those that are replacing worn-out or obsolescent sets. With TV sets lasting perhaps 8–10 years (particularly in future with ever-increasing reliability) this means a market of perhaps 2 million sets a year. Of this total, UK manufacturers might take less than half if cheap imports increased for any reason. When this is compared with peak production in 1973 of over 4 million sets, the word 'declining' seems not quite so loony. It is this that is the principal spur for the TV industry to move into new products.

Manufacturing and rental retail supply

The TV industry in the UK is unique in having a large and vigorous rental industry alongside the retail companies. Manufacturing in the UK means four giants and some others. The giants, strictly in alphabetical order, are GEC, ITT, Philips, and Thorn. Also in there are Decca, Rank Radio, Rediffusion, and sundry others. Of the big four, two are UK-centred (GEC, Thorn) and two are part of multinationals originating overseas (ITT from the USA, Philips from the Netherlands). However, all four qualify as multinationals and all have a substantial stake in UK manufacturing. In so far as TV production has a future in the UK, these giants will have a major say in it.

The rental and retail industry is very mixed. At one end are the giant rental companies such as (again in alphabetical order) DER, Granada, Radio Rentals, Rediffusion, Visionhire. Also at this end are the major retail companies such as Comet Warehouses, Currys, Dixons. Major department-store chains such as Debenhams, the Co-op, Lewis's are significant TV outlets. At the other end is, almost literally, the corner shop. There are still a large number of small retailers with TV sets as part of their stock and who emphasize their 'small is beautiful' appeal. The buying power of the major rental and retail companies means a powerful 'pull' and 'push' on the manufacturers. The market suppliers are at the sharp end and they know only too well what the market is doing. With experience, and a little luck, predictions can be made about the future. These predictions can mean pressure on manufacturers to supply more or to cut back. To work well, the relationship has got to be a close one. For this reason, the manufacturing and supply industries have moved ever closer. This relationship can be explored through two contrasting cases.

The Thorn Group has a wide variety of TV-related interests. It includes a manufacturing company, a major retail chain (Rumbelows), and three rental groups (DER, Multi-Broadcast, and Radio Rentals). Its rental outlets specialize in Thorn products (e.g., Ferguson) but not exclusively. Its retail outlets feature many products. This broad-based integrated approach provides strengths and weaknesses. There is integration of manufacturing and supply, but a heavy reliance on internal products.

The Granada Group (for which the writer works) has a quite different approach. Granada TV Rental claims to be the world's largest independent TV rental company. That advertising slogan is not always understood because the crucial word is 'independent'. Granada has no manufacturing interests (in the investment sense) and pursues a policy of independent purchasing. In principle, Granada can purchase from anyone, anywhere. The practice is necessarily less volatile. Two of the four giants, GEC and ITT, are prime suppliers to Granada and this relationship is probably no different, in day-to-day essence, from between any of the main rental companies and their manufacturers. Continuity of supply is vital, changes of units (with all the implications on servicing, spare parts, etc.) are major decisions. So, generally speaking, the major specialist rental companies tend to feature the same manufacturers' products over quite a long period.

83

In the retail trade, where servicing is not necessarily offered to customers long-term, it is easier to change suppliers. This tends to work to the advantage of the customer in providing a healthy variety of goods, but the UK TV industry has rarely been the beneficiary. As in many other industries, short-term domestic production problems often cause retailers to find alternative sources of supply abroad. At present, the whole industry faces increasing difficulties. The total TV market is approaching saturation in terms of one colour set per household. 'Second set' sales, whether purely for home use or for portability, are in the main going to overseas manufacturers. The rental industry faces lower growth as increasing reliability and lower prices cause the customers to switch to retail. And the retail industry is having to fight the same constant battle against overheads and competition as every other area of the trade.

The TV industry and viewdata

Perhaps there has not been an actual presentation of a viewdata 'wooden spoon', but there is more than a little suggestion from some quarters that the TV set manufacturers are a bunch of hesitant, shiftless layabouts because of their lengthy hesitation to start mass production of viewdata sets. The manufacturers have a different tale to tell.

Let us go back in time to the days when Sam Fedida, inventor of viewdata, was a name known only in the households of the staff who worked for him. By 1973, the Post Office Research and Development Laboratories had been looking at telecommunications and television technology for some time. The principal reason related to viewphone—for seeing as well as chatting to the other person. But technical and cost problems consigned that idea to the shelf. However, the work provided a spur to the somewhat different project of linking telecommunications and television.

In three quite separate research areas, work was proceeding which was to come together dramatically as viewdata. In the first place, Sam Fedida and his Post Office colleagues moved to the operation of a small information system using a Hewlett Packard computer. The television sets being used to display the information were in 'mono' and the screen format was 32 characters per line and 13 lines per screen. This was the origin of viewdata. The second area of research was being undertaken by BBC engineers to utilize spare capacity on the broadcast TV signals. This research had led to the development of Ceefax ('See Facts') as an information distribution system using ordinary TV sets with suitable modifications. This work had resulted in a close collaboration between TV industry research laboratories and the BBC to develop the world's first teletext service (as this type of system became generally known). The third area of research was bringing together microelectronics and telecommunications. The transmission of computer data had grown considerably since the first telephone data services had started in the UK in the mid-1960s. Data transmission units were large and clumsy and the search was on for ways of reducing these units in both size and cost. Microelectronics yet again was the key.

So now we had all the ingredients: an embryo viewdata system; development work to turn TV sets into information terminals; and low-cost telecommunications to provide the transmission system. The ingredients came together with contact between the Post Office and the BBC on Ceefax and viewdata; between the BBC and set manufacturers on Ceefax; between the Post Office and the telecommunications industry on low-cost viewdata components. The contact rapidly developed a common standard for teletext (Ceefax and the IBA's teletext system Oracle) and viewdata with 40 characters per line and 24 lines per screen. The TV industry was now faced with the development of the electronics needed to drive both systems. As a number of elements are common between the systems, this approach obviously made sense. The result of this uniquely collaborative venture was a modified colour television set containing a teletext and viewdata decoder to produce all three information services, Ceefax (Ceefax 1 on BBC 1 and Ceefax 2 on BBC 2), Oracle on ITV, and the Post Office's viewdata service. At this early stage there were a number of external units associated with viewdata. The telephone had to be used to call the computer centre, and the signals needed converting between TV and telephone line by a Post Office modem. Development towards a fully integrated set of electoronics depended on another industry—the microelectronics manufacturers, often known in the jargon as the semiconductor industry (a transistor is one 'semiconductor').

The development of viewdata involves three novel areas: the computer programs, the information, and the TV set. While all three areas can claim a degree of pioneering, the only real technological novelty is in the TV set. Initially the only unit being built into the sets was the decoder, and this was so large, using standard electronic components, that only 26-in (66 cm) colour TV sets could accommodate it. But the Post Office decided in early 1977 to allow manufacturers to build in the modem and other components, thus opening up new production possibilities. From the beginning, a number of companies took a very committed approach. GEC Semi-Conductors, General Instruments (GIM), Mullard, and Texas Instruments took a significant interest, this stemming partly from teletext developments. All four of these major semiconductor companies worked closely with the Post Office and the TV industry to produce viewdata components. The TV industry itself agreed to supply over a thousand sets for the viewdata market trials. In addition, a number of telecommunications suppliers also entered the lists and offered sets. The net result, in response to a request for a thousand sets, was a promise of over fifteen hundred units from over a dozen sources. Delivery was scheduled between June and December 1978.

In fact, as is now well known, actual deliveries were dismal, at least compared with expectations raised among Post Office officials and the information providers. So what went wrong? The fundamental problems are within the process of developing microchip technology. The building of circuitry with discreet components is relatively expensive but when something goes wrong, it is possible to identify and replace components. The creation of microchips containing the equivalent of hundreds or even thousands of components is immeasurably more complex—but it is much easier to produce thousands of them at low cost. So the stage which has to be completed successfully is design and initial production. Once this is successfully completed, however, the required units can be mass produced in tremendous numbers, giving a low-cost powerful unit.

But the difficulties are also considerable. The reduction in size of conventional items can be very tricky. Miniature pieces of silicon do not behave excatly like the components they replace and there is no absolute certainty that a circuit which works perfectly in one type of component will do so in another. This problem has been a major factor in the late development of viewdata electronics. For example, before 1977, no modem had been built in miniature form which would work within a TV set. The theory was all there but the practice was not. And yet, building on the considerable laboratory work undertaken within telecommunications, during 1978 several manufacturers produced working devices able to operate inside TV sets. The major task remained mass production. It is really at this stage that the commercial growth of viewdata moves into the classical 'chicken and egg' dilemma. During the build-up in viewdata interest, each of the parties concerned had to make its own predictions of the likely market. The Post Office saw in viewdata a means for increasing the number of telephone calls made by residential customers. The bulk of the telephone service is provided to service the needs of the business population. During the day this equipment is heavily used but in the evenings and at week-ends much of it lies idle.

The Post Office therefore wanted to see a mass residential market for viewdata within a few years, but has now moved to a more pragmatic view. This view is broadly shared within the viewdata world that amounts to 'business early, residential later'. What is in question is how soon viewdata will become mass-market. If it is accepted that there is a short-term business market measured perhaps in hundreds of thousands within five years, what is the sensible policy for the TV industry and their suppliers, the semiconductor industry? Strategies can be aggressive or passive. The aggressive approach involves what might be called 'market push'. This involves a commitment to the product and heavy marketing to persuade customers to buy. This passive approach could be called 'market pull'. The industry responds to demands from the marketplace for products seen by the customer to be needed. Either way, expert marketing is needed but the perceived risk is different.

The semiconductor industry, the TV manufacturers, and the end suppliers, rental and retail, need to make common cause for production to rise. After all, it is no use a semiconductor manufacturer producing a new viewdata module (at great expense) if the TV manufacturers do not want to buy it. Equally there is little value in TV manufacturers setting up a production line if there are no orders coming in from the market suppliers. So who takes the risk?

The fundamental point which is sometimes overlooked by the TV industry in dealing with viewdata is the crucial role of the information. It is, after all, no use trying to sell an information service that does not exist. Th confidence of the TV industry generally is therefore seriously affected by the quality of the database. It must be said that only a relatively short time ago, the database on Prestel, as the Post Office viewdata service had come to be called, was not particularly good.

During 1979 the situation improved immeasurably and the committed Prestel information providers must now feel they deserve the full support of the TV industry. It is now fairly clear that viewdata has taken a firm grip on the business world. Certain industries have embraced Prestel vigorously and others will follow. This has acted as a clear green light to the major committed members of the TV industry that viewdata is really moving. For the first time, a number of TV manufacturers have set up small production lines to produce viewdata sets. The semiconductor manufacturers can now see the beginnings of the real market growth necessary for their investment to be repaid. And vigorous competition in the market place for the customer's business is getting under way.

There is still a long way to go to the Post Office vision of a million Prestel subscribers, but at least the numbers are now growing. The TV industry as a whole may still to some minds deserve the wooden spoon, but there is no lack of commitment any more. What does need to be placed in perspective, however, is the place of viewdata in the economy. At the time of writing, the UK economy faces a downturn. Inevitably this will place pressure on companies to cut back on expenditure and viewdata may well suffer a setback as a result, even though it can be argued that businesses under pressure need up-to-date and accurate information even more than usual. The domestic market for Prestel cannot be helped by the general economic situation either. As well as competing against other consumer products, the basic question remains, 'Is it within economic reach anyway?' In deciding how much to invest in domestic viewdata products, the TV industry also has to recognize its own internal competition for the customers' money. On the rental side, teletext and video recorders are already an alternative attraction for customers. On the retail side, hi-fi and radio are just two more side-attractions. There is as yet no obvious sign that domestic viewdata, and the mass volume so essential to microchip low-cost technology, is less than three or four years away from overtaking business viewdata in volume. And during that time we will be seeing such wonders as video discs to tempt money in other directions. So the TV industry and its component suppliers will clearly invest accordingly.

Viewdata was invented in the UK and is further advanced in commercial terms in the UK than anywhere in the world. A question hanging over Prestel is whether the UK will pay the price for pioneering. The Post Office has succeeded in selling its computer software to a number of countries abroad and this obviously helps the general cause of standardization. However, it would be unfortunate, to say the least, if UK viewdata manufacturers found that the export market was lost because overseas standards were not Prestel-compatible. Until the overseas markets mature, UK investment on this score will necessarily be limited.. As the markets do mature, and as international standardization develops, there will obviously be corresponding growth in the overseas indigenous TV manufacturing industries. Nevertheless, the major UK TV companies are already playing a significant role in international viewdata which can only boost the growth of UK manufacturing capability. It has been said that over half the new jobs created in the USA during the 1980s will be in small companies with less than 50 employees. This 'cottage industry' view is related in part to the advent of the microchip and the proliferation of small companies to serve this new technology. So unlike the microchip manufacturer who was so successful that he had to move into smaller premises, the microchip service industry looks set for a firm period of growth.

Already, with a considerable amount of government backing, small companies are producing viewdata-related products and services from intelligent terminals to specialized consultancy. The activities even of the big companies are also still largely centred on a relatively small group of staff, but with the emphasis moving firmly towards mass-production. The small manufacturers have the versatility to supply specialized units to businesses as part of the process of bringing viewdata into every organization. In some cases these small manufacturers, who are almost entirely in the electronics and computer industries rather than in the TV industry, will provide a spur to the TV industry in its search for new applications and new products.

In the short term, the continuing technical difficulties being experienced by the semiconductor industry will be an irritation rather than a major drawback. Production will not be as high as other committed viewdata organizations would like. Investment will be modest as the economic situation and general uncertainty prevail. There will, however, be increasing efforts by the committed rental and retail companies to develop viewdata in the business market and in the embryo domestic market.

In the longer term, the TV industry has to invest in new products to survive. Viewdata will face severe competition from video, both tape and disc, as well as other consumer products. Whether it achieves its true potential within the TV industry is unclear. Its potential, however, is immense and it is only by a major commitment to mass production that this potential will be realized. The timing, however, is crucial and it needs to be remembered that the S-curve, of growth can be a long one. Television took six years after the Second World War to reach half a million sets; colour television had been operating in the USA for over fifteen years before the UK reached half a million colour sets; home video recorders have been around for over a decade but there are still only about 140 000 in use.

The interrelationship of computer, business equipment, telecommunications, and TV industries in the business world, fed by the crucial semiconductor industry with its microchip production, will at the same time create new opportunities and new challenges for viewdata. There is much uncertainty hanging over the TV industry and its future: viewdata holds out a promise that the industry cannot ignore.

Part Three

Case studies in electronic publishing

Editor's introduction
Why become an electronic publisher?

Motives

What are the motives to become an 'electronic publisher'?

Many of these apply to all information providers to Prestel and to other viewdata systems. But they have especial force for a 'traditional' publisher of newspapers, magazines, etc., and even more so for a publisher of business information.

1. A basic instinct that much business and financial information is being computerized, and that computer-based distribution systems will increasingly be a bread-and-butter means whereby business people, not to mention civil servants and others in the office context, will get their information. Whether this will in due course cut into sales of traditional paper-based products, or whether it will be an addition to it, as business life gets more complex and demanding, is a matter for argument. It is noticeable, for example, that over the past two years or so the statistical content of the *Financial Times*—that is, material which most readily transfers to the computer—has been considerably increased in response to reader demand. It may be that there is a general process going on, a demand for more and more accurate data, which will be fulfilled in a number of ways. But obviously, one must also be prepared in the long term for some impact on sales of paper products.

2. However this process works out, it is clear that electronic publishing, however defined, offers a clear chance for business diversification, for business and other publishers. This is a simple commercial opportunity, which might have been carrier pigeons or smoke signals, but just happens to be electronic transmission of data. It is simply a new market opening up.

3. By the same logic, it is clear that if one company does not step in and exploit this opportunity, others will. Market forces operate here as elsewhere, and there must therefore be a defensive element as well as an offensive one. It should be remembered that there is no prior definition of who can, and cannot, be an electronic publisher. New entrants, or entrants from other walks of life, enter the arena. The landscape looks different from paper publishing, and quite rightly, it is a chance for entrepreneurial challenges to arise.

4. For a newspaper company, it is a chance to explore the potential of new technology at arm's length from, though not totally away from, the familiar restrictions of the immediate newspaper environment. Viewdata represents a useful opportunity to keep pace with computer systems by other means.

5. Viewdata is also a means and opportunity for exploring new combinations of people and skills. Like many other industries, the newspaper industry is fairly tightly demarcated between unionized functions, e.g., journalists, compositors, clerical staff, library staff. Viewdata offers an opportunity to mix together people from different background into one, single team, with cross-fertilization of skills and attitudes. It has been one of the most rewarding aspects of viewdata to see how people, free of formal restrictions on what they can do, branch out and develop in unexpected ways.

6. Viewdata systems enable their participants to explore and evaluate the concept of the database. This is a much wider concept than viewdata itself, and is applicable to many computer systems. Many newspaper and other publishing companies are moving into this field, seeing themselves more and more as 'information companies': but it can be expensive, and is certainly a novel, even foreign territory to those more familiar with year-books or late editions. Viewdata is an easy, relatively cheap and accessible way of exploring the database concept with a view to future decisions on investment, outlets, commercial policy in this new but developing field.

7. A further motive is to explore the relationship between viewdata systems and other electronic systems. As more and more activities, from letter writing to typesetting, from business forecasting to games, fall under the spell of the computer, so the relative role of viewdata will develop and change. Is it, for example, basically a vehicle for data held elsewhere; a distribution and marketing tool? Or is it, in addition, a vehicle for original material, or an original repository from which other computer systems can draw? In short, what is viewdata's place in the general electronic environment? An electronic publisher needs to know.

8. By becoming involved in viewdata early in the game, a publisher has a chance to influence the development of the medium in ways favourable to himself, and to the commercial future of the medium as he sees it. Traditional publishing disciplines and constraints do apply, *mutatis mutandis*, to electronic publishing as well.

9. Early participation is also a business opportunity to set up, or try to set up, a viewdata 'hot shop', with an all-round capacity in viewdata matters. There is some reason to suppose that the market will only sustain a limited number of these specialist viewdata companies offering a complete range of viewdata services (defined below), and that it may be the early birds that catch this particular worm. While viewdata is too young, and changing too quickly, to make dogmatic statements, it is to be doubted whether the idea of every Prestel user being an information provider, and vice versa, will ever be more than partially realized. As with paper publishing (which in theory any organization or individual can do), there will tend to emerge in practice a limited group of specialist companies with the skills, people, premises, and experience to handle a wide range of products on behalf of other people.

This is not a complete list of motives: not all aply to everyone. Each publisher has to make his own decision whether the balance lies for or against participation in viewdata.

Components

What then are the components of an electronic publishing business in the full sense?

Again, these will vary from case to case, but the components might be as follows:

1. An open mind about what to publish in the new medium. Clearly, your existing data must be your starting point. But too many newspaper, magazine, and book publishers see viewdata as some sort of 'electronic clone' of their original paper publication, or as a poor man's version of it. Worse, they see viewdata just as a trailer or taste for the 'main' publication, or as an advertisement for it. This is not only excessively defensive and unimaginative: it is also courting disaster, because the Prestel/viewdata audience will surely not accept in the long run second-hand, derivative, cut-down versions of what may anyway be cheaper to buy elsewhere. Whatever you publish on viewdata has to be right and convincing in viewdata terms: it may in many cases be very different from what your home publications look like. In short, it is a new publication, to be looked at with a fresh mind.

2. At this early stage, it may also imply a diversified viewdata publication that is flexible and to a degree disposable. No one can yet pretend to have found the right product or product mix for viewdata. Therefore it is necessary to hedge one's bets, try out alternatives, with ways of measuring the success or failure of those alternatives. Again an open but experimental mind is called for.

3. Just as the database may be diversified, so should the people be diverse. It is still too early to say whether the journalist's eye, the computer person's eye, the librarian's eye, the graphic artist's eye, the book publisher's eye, or any other eye, will perceive the best uses of viewdata—or whether it will be a pooling of all those and other perceptions. Again, it is necessary to hedge one's bets.

4. By a similar logic, the outlets for a specialist viewdata company must be diverse. For a UK company, of course the Prestel system must be the primary focus of interest, especially at this stage of viewdata development. But viewdata promises to be a world-wide phenomenon, in which national boundaries become increasingly meaningless as the technology, and international standards for it, are evolved. Some data is very parochial: but a surprising amount is not, provided that the person in the other country can have easy and cheap access to it. The world is your oyster as a viewdata publisher, far more immediately and dramatically than with paper publications, even allowing for language barriers.

5. But the most important component is what results from these diversified people, outlets and attitudes, and that is a diverse range of activities that go to make an all-round specialist viewdata publishing business. This range of activities may clearly vary from case to case, but might include the following:

93

- simple information services on Prestel or other viewdata systems: in other words, the original information provider role, which is probably the cornerstone for the rest
- advertising and marketing services that march alongside the main information services, and are offered, as are all advertising outlets, to third parties such as companies, public organizations, etc., to make use of for a fee
- contract work for others, taking their data and converting it into viewdata frames, for a price
- consultancy, advising others on what is, and is not, possible on viewdata systems given the nature of their original material
- technical development work, devising 'add-on' systems that answer to the information handling needs of particular IPs, oneself included
- conferences and seminars, both to make money and to propagate the viewdata message
- market research, both one's own, and for others on a contract basis
- sale of data frames to other, perhaps overseas, IP's for use on other viewdata systems. Being already in viewdata format, such material is readily recorded and transferred, at a negotiated price
- help with, and contract work for, closed user groups, private viewdata systems, etc.
- becoming a direct IP to viewdata systems in other countries, and offering contract services for, and to, those other countries also.

Limitations

Having therefore become an electronic publisher, what are the broad considerations that govern—or limit—the type of information service that may be offered on viewdata?

These may be summarized as follows:

1. The viewdata screen is a narrow window on a large world. The screen size has severe implications for editing techniques: the narrow-window effect has strong implications for indexing, routing, database structure, and other means for guiding the user round the information offered. Unless the material is susceptible both to the editing techniques implied, and to the database structure implied, then it is not suitable for viewdata.

2. Viewdata is a mixed medium of text, figures, and graphics. This makes it unusual, if not indeed unique, both in the world of television and in the world of computers. Maximizing the use of these facilities, and finding the most judicious mix between them, is the art of the professional viewdata editor. Clearly the mix may vary from IP to IP, but the aim must be to exploit the unique advantages of a viewdata system, of which this range of display possibilities is the most important.

3. The ability to put a price on each separate page of information is also a unique feature, both in TV terms and computer terms. I refer to it as 'unbundling' in the computer sense. Most products in publishing are bought

for some one-off sum, a cover price or annual subscription or, in the case of BBC TV, an annual licence fee. But viewdata is normally bought item by item, page by page, and the price can vary from page to page, and can be changed for any given page. There is therefore a flexibility of pricing—and purchasing—especially given the facility to go off into a closed user group with yet other pricing methods, that is unknown elsewhere.

4. There is, as already mentioned, a quite new competitive environment. The fact that you are an established publisher of a certain type of information on paper, while it puts you in a strong position, does not guarantee you a similar position on viewdata, simply because the different medium opens up the possibility of others challenging your position. This in itself can be a strong motive towards putting up certain types of material, to establish a position, always remembering that the competition may come from unexpected, even unwelcome sources.

5. One fundamental conceptual problem is whether viewdata is most suitable for highly volatile information, such as today's news and the latest market prices, or for relatively static information, such as train times, restaurant guides, etc. A related question is whether the commercial objective should be to have a smaller number of pages that change frequently,or a larger number of pages that change infrequently, if at all. Maybe viewdata can do both. If so, it will show remarkable versatility. The fact that pages are not automatically updated on the screen means that there is a limitation on viewdata's ability to be a 'hot' news medium: compare, for example, the London Stock Exchange TOPIC system, which is a viewdata system but has automatic continuous update.

6. Is viewdata a medium for hard facts only? Figures, dates, times, addresses, prices. Or is it also a medium for debate, opinion, argument, controversy?

    The limitations of the screen size mean that viewdata is not suitable for extended amounts of text. You cannot put a novel on to viewdata, nor yet a full-length feature in the magazine sense. So if viewdata is a medium for opinion, they have to be succinct opinions, statements of position, rather than reasoned cases. The whole question of whether people will genuinely *read* viewdata, as opposed to taking in salient figures and words at one glance at each screen page, is an important one and difficult to decide until people have got more used to using the TV screen for these types of purposes. But what it certainly emphasizes is the point made previously, that a viewdata information service cannot be regarded as some sort of 'electronic clone' of the printed publication. The style, usage pattern, and economics are quite different. Another way of stating the question is whether people will browse through viewdata, or go in to find a specific bit of information and either get it or not get, in a hit-or-miss manner.

7. And what is the role of advertising and promotional material of other sorts?

    Advertising itself does not fit traditional notions or patterns when applied to viewdata. But most traditional publishing, not to mention television, is heavily subsidized by advertising revenue. Can a similar subsidy be expected for viewdata? How big a proportion of the information that people view can be paid for, sponsored, by advertising of some sort, without these people being 'turned off' the medium? If it makes the information free, or nearly so, perhaps the answer is, 100 per cent, provided that the suppliers use their common sense about the credibility and reliability of the information they put up. But perhaps there will be, must be, information services that are 'pure' editorial, paid for at full cost by the user, especially the business user. It will be interesting to see the balance that emerges between the two approaches to the philosophy and economics of electronic publishing.

8. Most of these considerations will get a different answer depending on whether viewdata is successful in the business market or in the residential market, or in both: and will get different answer from a business IP compared with a residential IP. Broadly, business is used to paying for information as a resource, but the man and woman in the home are not: people in business are used to searching for information as a skill, but the man and woman in the home are not. Arguably, the most natural technical application of viewdata is to the mass market, given that TV sets and telephones are mass-market objects (even though, in the case of the telephone, not everyone in the UK has one by any means); but the economics and user skills required for viewdata may push it more in the specialist and/or business direction.

9. Lastly, a case can be made out for saying that viewdata will only take off in a big way, with high usage, when the buying and selling mechanism via the response frame, and the message facility, turn it into a much more active and novel medium, rather than the simple information medium that it basically is at present. Perhaps the information content will be the marzipan on the cake, the basic cake being the quite new and exciting ways that people can communicate with each other, and do their shopping, travel bookings and so on.

A note on costs

Will it make money? Nobody knows for sure, simply because no values can yet be attached to the average pattern of usage of the average business or domestic customer. Crude numbers of viewdata receivers in the market place tell you very little. They may be used frequently, infrequently, or not at all. Nor do we yet know the price sensitivity of the information displayed, or what, if any, competitive constraints there will be on pricing, either from rival IPs or from rival media. For any given page, we do not yet know at what price level the customer will refuse to look at it, or at what level someone else will step in (computer capacity permitting) and undercut.

However, some at least of the basic building blocks of costs and revenues on Prestel have been fixed, or can be estimated. Let us first deal with the tangible costs to the customer. These costs are quite complicated to work out, because the user of Prestel pays both the TV manufacturer, the Post Office, and the Information Provider for use of the system. To the TV manufacturer (or rental company) he pays at the moment up to £1000 purchase price, or £30 per month rental. These figures are variable and real prices will fall when volume production increases. To the Post Office he pays, first, the local telephone call charge: this money goes to the Post Office telecommunications department. Secondly, he pays a Prestel usage charge; this money goes to the Post Office's Prestel department, a separate profit centre.

Then, the user has to pay for the information itself. The system at present allows the IP to charge between 0 and 50p per page. The pricing pattern emerging from IPs is for advertisement and routing pages to be priced at 0p; for residential and consumer information to be priced at 1 or 2p per page; and for business information to be priced at 5, 10, or 15p per page.

From these figures, we can make our first rough calculations. Assuming that it takes 15 seconds for a Prestel page to be displayed and read and that a user will look at four routing pages before reaching an information page, the average total cost to a user for each page of useful information in the residential database, at off-peak rates, will be about 3p; and in the database, at peak rates, about 12p.

The IP also incurs a variety of costs for the risk or privilege of contributing to Prestel. First, the Post Office charges an 'entry fee' of £4000, and £4 per page per annum for each Prestel page booked. This means that an IP with a significant database—at least 1000 pages— will have to spend £8000 to join the Prestel club.

On top of this basic contribution is the cost of running the IP's database. It includes the cost of terminals and the use of the telephone line (most updating and inputting is currently done on-line) which goes to the rental companies and the Post Office. It also includes normal administrative costs, especially labour costs, which vary widely depending upon the size of the database, the frequency of up-date and editorial standards. The consensus among large IPs is that the overall cost of inputting and maintaining a page on Prestel, including overheads, is at least £30 a year.

Many people outside the information industry regard this as a surprisingly high figure. But information is by its nature a highly labour-intensive business, and so far the indications are that Prestel will be no exception. As we shall see, with a large database, labour costs can dwarf any other costs: there seems to be a rough correlation among IPs of one editor per 1000 pages. Several large IPs have a total staff (marketing, management, etc., as well as editors) of 12 to 14 for a full viewdata unit at this stage.

In the light of these figures, let us examine how the economics of Prestel look to each of the parties involved—the Post Office, the set makers, the IPs, the customers—and how they diverge. For the Post Office, penetration of the mass residential market is regarded as essential if Prestel is to generate the revenue that will make it profitable. An additional appeal is that it gives an opportunity for a large, underutilized asset—the residential telephone network—to be more fully exploited. It may also offset falling employment in other areas of Post Office activity. However, the Post Office is primarily interested in ensuring that everyone with a telephone and a television set becomes a Prestel user. It is not particularly bothered whether the user buys a new TV set for £1000 or an adapter for his old set at £100.

For the television-set manufacturers, the mass residential market is the only one that makes economic sense, and then only if the majority of that market buys or rents a new Prestel-receiving set. The British TV manufacturers see Prestel as the possible answer to the flattening out of colour TV sales now that the colour market is nearly saturated. But the economics of mass production dictate that sets must be sold in hundreds of thousands before the cost of laying down a new production line (perhaps about £15 million) can be justified.

To get the price down to the level of normal-set-plus 15 per cent regarded as the level acceptable to buyers will itself require a market of considerable size. It is argued by the Post Office and the IPs that in order to create a mass market, the price of sets must be low to start with—in other words, that the manufacturers must start selling their sets now at mass-production prices. The manufacturers, understandably, want evidence that the mass market exists.

For the IP, the choice of what market to aim for is not so clear-cut. As indicated, the IP will need an income of perhaps £35 per frame to make participation in Prestel worth-while (some say more). This income can be generated by a lot of users looking at the IP's pages with a low price per page, or a small number of users paying a high price per page. It can also be generated by a lot of users looking at a few pages frequently or a lot of pages occasionally. Also, the IP is not particularly concerned whether the user has a new TV set, an adapter with an old TV, or even an adapted VDU, or whether the user calls Prestel on congested business or under-used residential telephone lines—just as long as he calls.

Here then are clear areas of divergence in the economic interests of the various parties involved. How serious this divergence is, and whether it is more powerful than the general interest in making the system a success, remains to be seen. There is no reason in principle why the divergences cannot be reconciled.

6. Who's who on the Prestel database

Ederyn Williams

In the final analysis, Prestel is only as valuable as the data it contains. The television sets and the computer would be meaningless without the 160 000 pages that feed and fill them. This fact has been recognized right from the start of the service and much of the basic policy of Prestel has been formulated with the aim of achieving the conditions that are necessary for a stable and vigorous information-provision industry.

IPs and sub-IPs

By and large, this policy has been successful: at February 1980 there were 131 information providers (IPs), with another 116 organizations acting as sub-IPs. A sub-IP has no contract directly with Post Office Prestel, but rather has obtained space from an IP in which his information appears. Normally the names of both the IP and sub-IP appear on the pages. The IPs acting in this capacity, known colloquially as umbrella IPs, retain legal and editorial responsibility for their sub-IPs. One can distinguish at least four types of relationships between IPs and their clients:

1. A sub-IP acts as a wholly independent supplier of information retaining most of the editorial responsibility and even the physical act of editing. Such sub-IPs would often wish to be a full IP if the Post Office could offer the database space, which it finds difficult at present. An example of this relationship is the House of Commons (sub-IP) with the Central Office of Information (IP) database.
2. An organization may merely make its information available to an IP, sometimes on a royalty basis, but more usually not (since information provision is not yet a particularly profitable business, because the number of users is still relatively small). An example of this was Gallup Polls (sub-IP) with the Applied Viewdata Services (IP) database.
3. An organization may choose to advertise through an IP. This is becoming increasingly common, and is a useful source of revenue to IPs. An example is Barratt Homes (sub-IP) with the Viewtel 202 (IP) database.
4. Individuals may author 'articles' for Prestel, again usually on a royalty basis. An example of this is Marion Ellis Cookbook (sub-IP) with Mills and Allen's (IP) database.

All these types of relationship are sprouting up in profusion, and adding to the richness and variety of the Prestel database.

Who are the IPs?

The answer to this question is—many and varied. The 131 IPs at February 1980 fall into the following categories:

1. National newspaper groups (four in total), *The Times*, the *Financial Times*, the *Express*, and the *Daily Telegraph*. Apart from the *Financial Times*, these IPs have been particularly inactive, primarily because they seem to be beset by who-does-what arguments with their workforces, which are delaying the introduction of most new technology. Unlike the situation developing in some other countries, UK national newspaper groups in general have not been playing an important part in developing the Prestel database.

2. Local newspaper groups (five in total), such as Eastern Counties Newspapers, The Birmingham Post and Mail, and Scotsman Publications. These IPs have generally been extremely active and capable, in marked contrast to the national press. Their ability to rapidly introduce Prestel into their organizations is in line with the ability that they earlier showed to introduce computer typesetting.

3. Magazine groups (six in total), such as IPC, Morgan Grampian, and the National Magazine Company. These have also shown themselves able to readily adapt to Prestel. The wide range of potential material available to them, from their various specialist publications, must help, as does their ready familiarity with the editorial and economic problems of small-circulation magazines—which is what Prestel most closely compares to at present.

4. Other publishing groups (fifteen in total) such as the British Printing Corporation, ABC Guides, and Link House. These groups seem to have found it less easy to adjust to Prestel, usually because they have some specialist types of publication which may prove difficult to convert into Prestel, or may prove less popular on Prestel than in their original format. Sometimes, lacking the wide range of publications of the magazine groups, this has been a real impediment. Nonetheless, several IPs from this category are among the most popular on Prestel.

5. Central government departments (eight in total). Among these are the Central Office of Information, the Ministry of Agriculture, Fisheries and Food, the Department of Industry, and the Central Statistical Office. Activity in the government sector has been commendable, though the fact that the IPs, and thus their databases, are explicitly mission-orientated, unable to change their business as they receive feedback from the market place, has meant that their databases are not among the most popular. Nonetheless, several of them are providing well-organized, well-presented, and useful services.

6. Government agencies (seventeen in total) including the British Library, The British Tourist Authority, the Meteorological Office, and Professional Executive Recruitment. Although all these agencies are mission-orientated, and thus limited in the range of topics they can cover in their database, several of them have adopted very successfully to Prestel. This may be

because, in some cases, their traditional role involves extensive communication with the public.

7. Nationalized industries (seven in total), including British Airways, Sealink, British Rail, and Post Office Telecommunications. Four of these are Post Office departments. Most are relatively successful, essentially for the same reasons as the government agencies.

8. Companies in other businesses (twenty-eight in total) including Cooks, Barclays Bank, Debenhams, Grand Metropolitan, Norwich Union, and Quantas. Although the largest group of IPs, they are far from the most successful: only three or four appear amongst the 30 most popular IPs. There are various reasons for this. Their material, which could be loosely described as advertising, is usually prepared by the company themselves: their inexperience in the media is frequently apparent. In addition, they usually have small databases, which, in some cases, are relatively infrequently updated. It is now more frequently the case that such companies go to an 'umbrella IP', who acts like an advertising agency for the sub-IP.

9. New companies (eight in total), such as Fintel (through which the *Financial Times* is involved), Teleview, Intext, and Network Data. It is gratifying to see that Prestel has stimulated the formation of several new companies, dedicated to exploiting the business opportunities offered. Nearly all seem to be doing well, regularly being among the 30 most popular databases. The reason seems to be that they are profit-orientated, so they can enter any or every aspect of Prestel that seems worth while.

10. Companies expanding from non-print media (four in total), namely Mills and Allen, Extel, Datastream, and Reuters. The first three of these have already shown themselves to be among the most popular of IPs, while Reuters has, at the time of writing, been unable to start a full service due to continuing internal labour negotiations.

11. Associations (nine in total) including the Consumers' Association, the Printing Industry Research Association, and the Institution of Electrical Engineers. This group has had mixed fortunes: the Consumers' Association is consistently among the most popular IPs, but some other associations have had such a limited and specialist database that it is hardly surprising that they are not very popular.

12. Computer software companies (five in total) including Baric, Aregon, and Langtons. Although they lack the publishing experience, these companies have tended to adapt quite easily to Prestel, presumably because they are able to put their computer experience to good use.

13. Miscellaneous (fifteen in total). These are mainly companies offering specialist information services or scientific consultancy, local authorities, and educational establishments. None are among the more popular databases.

As can be seen, the IPs are many and varied, both in type and in their success. We expect this variety to continue, though several of the smaller IPs may become sub-IPs under an umbrella.

Popularity

We can see which databases are the most popular by looking at the 'top ten' information providers. Our top ten, during the period 6 October 1978 (the start of the test service) to 28 December 1979, is shown in Table 6.1. What can be

Table 6.1 The Prestel top ten (March 1980)

Position	IP	Type	Information
1	Baric	Software house. Joint Barclays Bank/ ICL	Games, company information, theatre guide, travel, holidays
2	Viewtel 202	Local newspaper group. *Birmingham Post and Mail*	National news, sports, display advertising, 'What's On'
3	Eastel	Local newspaper group. Eastern Counties Newspapers	Games, quizzes, classified advertisements, local 'What's On', local travel, top ten records, jokes
4	Family Living	Magazine group. National Magazine Company	Advice, cars, games, quizzes, children's club, going out, restaurants, horoscopes
5	Sealink	Travel organization, car and passenger ferries	Ferry timetables, holidays, day-trips, rail–sea services, information for travel agents
6	Fintel	New company. Joint *Financial Times/* Extel	Business news, company information, economic and financial statistics, business games, display advertising
7	Which?	Association. Consumers' Association	Advice on holidays, buying a car, buying appliances, health, pregnancy
8	Mills and Allen	Media company	'What's On', legal advice, education, religion, reviews, display advertising, office equipment, motoring advice, cooking, stories, jokes
9	IPC	Publishing company. Reed International	Farming, electronics, boating, cars, cooking, games, video
10	British Rail	State railways	Train timetables, travel information, cancellations, fares, holidays, credit

concluded from this? Firstly, it is obvious that existing publishing groups are not dominating the scene, although several of the top ten do have some publishing interests. For the next ten, IPs 11 to 20 on the popularity listing, this trend is even more noticeable, with only two being paper publishers. Secondly, it is apparent that the most popular IPs are large and diversified. Clearly the larger and more varied a database, the greater the chance that at least some of the material will prove popular. Diversification is clearly an important means of ensuring survival during a period when users' reactions and preferences are still unknown. Thirdly, it is noticeable that most of the top ten are providing entertainment—games, quizzes, and jokes—as well as information. The popularity of entertainment has surprised many people, and it may be that the novelty value of Prestel is leading to play which will gradually fade away. However, present evidence shows no such tendency: perhaps people's need for entertainment is as great as their need for information. Fourthly, various information topics keep recurring. Separate monitoring of the Prestel index pages has confirmed that the following topics are particularly popular:

Games	Travel	Holidays
Quizzes	Business news	Company information
Stock Market	Restaurants	What's on
Sports	National news	Consumer advice
Jokes	Horoscopes	Cars

In contrast, certain types of information are not as popular as initially expected. For example, the various directory listings on Prestel have proved rather unpopular. Similarly, small classified advertisements have tended to wither away. Why should this be? One can advance two reasons. Firstly, such material is difficult to maintain on Prestel. To date, most of Prestel database has been input by keyboarding on a simple terminal without word-processing facilities. Complex listings where single entries must be inserted or deleted with other entries shuffling up are thus very difficult to maintain. Thankfully, there have more recently been several developments in intelligent editors and off-line computer formatting which have reached the required degree of sophistication, which may lead to a resurgence of this type of material.

However, the second problem is less easily soluble. When a reader looks at a directory or small ads in print, his normal method of reading is to scan rapidly down the page until he reaches an item he wants, which he then marks, and either uses it or scans on for another item. Measurements of scanning rate by psychologists suggests that over 1000 items can be scanned per minute by the average adult. Prestel, however, has a small 'window'; scanning is thus seriously restricted to a maximum of, say, 10 pages of 10 items or 100 items per minute, which is one-tenth of the rate for the printed page. Prestel is thus poorly suited to material which is normally scanned; the only solution is a much more sophisticated search procedure applied to a much more carefully classified set of items. Designing this has not proved to be an easy problem to crack. Incidentally, the problem is equally acute with telephone directories, which in the printed form are normally scanned at great speed. The videotex telephone directory built by the French PTT is a remarkably poor substitute, since it only presents one item at a time to the user: unless the user knows the exact name and address of the person he wants, he would have to search by an incredibly slow one-by-one method. Most users are likely to find this too tedious to be tolerable.

Finding the frames

A good page that cannot be found is a bad page. We must not neglect that part of the Prestel database that has the function of leading users to the end pages. Indexing is becoming a considerable skill and, as illustrated earlier, may be the ultimate limit to the extent to which Prestel can be used for some types of data.

The original model of the Prestel index was of a simple, ten-fold tree, as in Fig. 6.1.

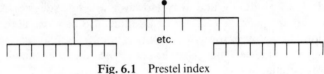

Fig. 6.1   Prestel index

Of course, it was quickly realized that using less than ten choices per page was advantageous, as this left room for expansion. Furthermore, it was realized that when users get to the bottom of the tree they must be given some way to get back up, or they are lost in a 'mineshaft'. Accordingly, people started building structures like the one shown in Fig. 6.2.

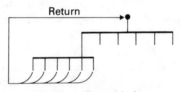

Fig. 6.2   Prestel index

The next level of sophistication was reached when it was realized that the pure hierarchy of a tree is a very poor match to the many and varied ways that people classify items. For example, 'cars for sale' could be classified under 'goods for sale' or 'motoring'; 'hotels' could be 'holidays' or 'accommodation'; the *Financial Times* Index could be 'stocks and shares' or 'financial statistics'. Given the lack of consensus, which was detected quite early by Post Office behavioural research, heavy cross-referencing is becoming the norm rather than the exception, giving trees as in Fig. 6.3.

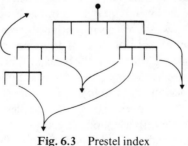

Fig. 6.3   Prestel index

Indeed, the 'trees' soon become too complex to draw on paper at all.

A further development was the idea of chained items. There are several types of items, such as news stories or monthly statistics, where a user may either want to quickly home-in on a single item, or to scan through them successively. These two different requirements can be accommodated in a structure as shown in Fig. 6.4.

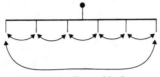

Fig. 6.4   Prestel index

This has also been called 'the wheel', since it can be drawn as in Fig. 6.5.

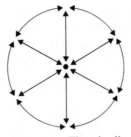

Fig. 6.5   'The wheel'

An even more sophisticated structure, nicknamed 'the lobster pot', has been described by Fintel, who have become one of the leading exponents of index structuring. This is illustrated in Fig. 6.6.

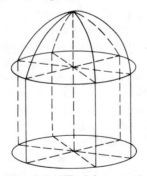

Fig. 6.6   'The lobster pot'

Two wheels are cross-linked to each other and to a main index page. This is especially useful for financial statistics, where a user may wish to look at just one figure, all the figures for one year, or all the January figures for several years. The more varied the requirements of users, the more complex the routing structures become—not that the user will perceive them as complex. Indeed, he should see such a structure as very simple to use, as helpful routing choices are offered to him at each stage, whatever his needs or interests.

Finally, some IPs are challenging the very division between index and information pages. They argue, quite rightly, that users find it tedious to go through a series of uninformative index pages. Since information pages must also direct the user on to other pages, or he is caught in a 'mineshaft', why not have every page contain both indexing and information, so that the whole database becomes a structured set of information pages. This concept has been applied most thoroughly to games, where the choices are themselves the content of the game. However, it cannot be stretched to include the uppermost levels of indexing, those run by the Post Office, where the topics covered by any one set of choices are so broad that it seems better to assume that the user merely wants to quickly rush through to a level which more concisely fits his information needs.

The future

How do we see the Prestel database evolving? At present, the overall form is still, at least in part, experimental, but every day the IPs are learning, through feedback from users, what types of material are popular and suited to the medium. Apart from an enormous amount of detailed change, the following trends can be expected to emerge.

Firstly, the public database seems likely to gradually lose some of the more esoteric specialist uses that can now be seen. Several applications are being tried out that seem to be more suited to private stand-alone viewdata systems. Companies wanting an in-house information system, or organizations offering information to a specific market sector such as stockbrokers or lawyers, may find that there are more gains than losses in having their own private system, supplied by Post Office Prestel or from elsewhere. For example, the London Stock Exchange is now planning their own viewdata system, called TOPIC, which will allow 2000 stockbrokers to have virtually continuous access to share prices.

One debate that frequently emerges concerns the use of Prestel for 'real' purposes (that is, for gathering information relevant to actual needs) versus the use for 'play' purposes (that is, for games, quizzes, jokes, pretty pictures, or serendipitous browsing). Initial opinions were that Prestel would be used almost entirely for serious purposes, and many people have been surprised, if not slightly shocked, at the extent to which play purposes have emerged. Looking at the history of other media suggests that this is not uncommon; certainly the inventors and early developers of the gramophone record, the radio, and television would all have been surprised at the extent to which these media are now used almost solely for entertainment. Although much of the use of Prestel is at present either for demonstration or is influenced by novelty, it is none the less true that early market research results suggest that the public interest in entertainment on Prestel is quite strong. So far, there have been no indications of the play uses fading away, nor do I expect such trends to develop. Play may not become the majority use of Prestel, but neither will it fade away to insignificance.

7. The electronic newspaper

Pat Montague

Viewtel 202 is the viewdata service of The Birmingham Post & Mail Ltd, operating through the British Post Office's Prestel service. Our introduction of the concept of an electronic newspaper on to Prestel stems from the earliest involvement of The Birmingham Post & Mail Ltd in the original viewdata pilot trial, organized by the British Post Office in late 1975. Our approach has not changed in principle since then, although it has in detail. To demonstrate and explain our view of how we will move forward in videotex, it is important to understand our origins as publishers, and also to examine our experience to date in Prestel. Our path forward is largely governed by our experience in traditional newspaper publishing and, more recently, in electronic publishing. This is combined with, we hope, a creative awareness of the extensive new opportunities that videotex offers to us.

The Birmingham Post & Mail Ltd operates the largest daily newspaper publishing centre outside the national press in England. The newspapers published are:

Newspaper title	ABC circulation July–Dec. 1979
The *Birmingham Evening Mail* series inc. *Sandwell Evening Mail* and *Sports Argus*	341 333
Birmingham Post	42 277
Sunday Mercury	195 815

The *Evening Mail* series has the largest ABC six-day evening paper circulation in the United Kingdom, and is third only to London's *Evening News* and *Evening Standard* on a Monday-to-Friday basis. The *Birmingham Post* is an up-market morning paper with a circulation of 42 277. It sells more copies within Birmingham than *The Times*, The *Guardian*, the *Financial Times*, or the *Telegraph*. The *Sunday Mercury* is the largest regional Sunday newspaper with a circulation of 195 815. It outsells virtually all of the eight national newspapers available within Birmingham.

Apart from these publishing interests, BPM Holdings Ltd, the financial holding company for the group, is also involved, through West Midlands Press Ltd, in ten weekly newspapers in the West Midlands area; has majority shareholdings in twelve London weeklies; has significant involvement in an evening-plus-weeklies group in North Wales; and has financial involvement in local commercial radio in the West Midlands.

Why, then, with these significant and prospering publishing interests in newspapers and commercial radio, is The Birmingham Post & Mail Ltd investing substantially in videotex? And, in particular, in Prestel? Why should it use as its vehicle the concept of 'the first electronic newspaper' and what does that mean?

Viewtel 202's origins lay in our fear in 1975/76 that electronic media, using television or VDU screens in the home and in business, could do us serious damage as newspaper publishers. Our experience at that time in the use of computers for commercial and text-handling operations had only served to increase our fears that totally new publishers could emerge, uninhibited by deep-routed trade union/management philosophies. Perhaps even more critical, these new publishers would need relatively low capital investment; negligible staff and overheads; no printing presses; no newsprint; no distribution costs; no traditional production or maintenance costs; a small number of editorial, sales and input/administrative staff; plus a relatively small computer operation. Not only this, but apparently the Post Office were creating a new, alternative medium for the masses, with an expectation of millions of suitably adapted TV sets in the market place within a decade, at low capital and usage costs. Utopia for the new, high-technology-based publisher. We viewed the new medium as being yet another competitor for people's weekly available cash. Our involvement with the British Post Office and the original group of information providers to the Prestel project was, therefore, largely based upon fear, combined with a fair-sized element of curiosity.

Until mid-1978 we kept our investment at a low level, with more management time invested than cash. We already knew by then that the scale of the Post Office proposals, combined with the lack of confidence by the set manufacturers in the intended rate of progress, meant that a genuine national public service, with large-volume set availability, would be well behind the original targets in time scales. Further, that the whole cost structure of the Prestel project was actively against it becoming a low-cost medium for information providers. Their charges had escalated four- or five-fold. Total user costs had escalated rapidly, too. In our view they were already prohibitive for a rapid development of the medium in the mass market. It was, and is, in danger of becoming business orientated, with no cohesive development of the data in the system to enable an attractive package to be easily and cheaply available to the whole mass of potential users. This pattern of a relatively high-cost business-orientated medium eased many of our concerns about Prestel's immediate impact upon newspapers.

The specialization of particular IPs; umbrella arrangements with indiscriminate groupings of sub-IPs; and the development of endless entertainment gimmicks in the form of quizzes and games, has led to a commercially dangerous development pattern in Prestel—it is not only expensive but it is also difficult to use for those wanting to access a broad package of needs and entertainments. This pattern of developments encouraged us, in October 1978, to launch our own trial package for the Motor Show 1978 at the National Exhibition Centre (NEC).

It took the form of the 'first electronic newspaper'. It set out, in a rough-and-ready way, to provide a broad-based package of useful and, hopefully, necessary information for people in Birmingham attending the Motor Show. The emphasis was upon minimum cost, and with maximum simplicity. This was to be our first public feedback of market-place information. With the help of British Relay and Radio Rentals, two sets were placed on Exhibition Stands with controlled usage. A further five sets were positioned at public-access points in hotels. All the sets were available for use for ten days. The accesses to the seven publicly available sets were a revelation—nearly 70 000 accesses in total, or nearly 1000 accesses per set per day. No one attended the sets to give advice or instruction—they were simply tuned to Viewtel 202 with no information on how to access other IPs' databases. No charges at all were made to the users (or the hotels). Unfortunately, there were no individual frame access figures available at that time, so we could only assume that our frames, as a package, were of interest to hotel visitors.

The topics covered were:

National news	10 stories updated 12 hours per day
International news	10 stories updated 12 hours per day
Motor Snow news	10 stories updated 12 hours per day
Business news	10 stories updated 12 hours per day
Financial news	10 stories updated 12 hours per day
Sport	As appropriate with maximum of 10 stories
Horoscopes	
Lucas Industries	Company profile
Lucas Industries	Programme of events for the Motor Show
'What's On'	Covering hotels, restaurants, night clubs, theatres, cinemas, sports fixtures and facilities, sauna and solaria, etc.
Train timetable	Cross-references from and to Viewtel 202

The basic conclusion drawn was that a simply structured information bank, with zero cost, had a high degree of public acceptability.

This experience encouraged us to initiate a small-scale regular news service, with one full-time journalist and occasional extra support. Inputting was handled by aptitude-tested compositors. Marketing was put in the hands of a senior member of the advertisement department, who was allocated full-time to Viewtel. The first electronic newspaper was launched. What has happened, then, since this initial launch?

From this initial low-key approach we have now arrived at specific objectives of creating a cohesive controlled information bank providing a national information service with a regional slant whenever relevant. This will be aimed at both the business and the mass residential user wherever possible. Games and quizzes are, and will continue to be, included—over 30 per cent of all accesses to Prestel at this time are in the entertainments categories so it would be foolish to ignore them. Our target market is probably as wide or wider than any other pure IP (i.e., one does not encourage sub-IPs other than in relevant topics). We are

109

also almost certainly less inhibited in the range of material we are willing to include to attract a wide base of users. Thus, the key areas that we see as opportunities are:

1. *Mass-market applications.* These will have broad-based market appeal, based upon relevant information, with good database control so that users can locate a wide range of relevant information simply and quickly. Apart from 'home entertainments', it is not intended to levy an IP frame charge for such information. Both business and residential appeal can be achieved if appropriate information is selected. This will be particularly appropriate to encourage the business-justified residential videotex TV set.

However, the majority of the mass-market users will be more likely to be paying their own telephone charges, computer access charges, and IP frame charges. Taking a broad spectrum of the information on Prestel, it is quite possible for frame charges to be twice the combined amount of call charges and computer access charges, outside 9 a.m.–6 p.m., Monday to Friday. Total 'information' costs for 10 minutes per day, 7 days per week usage of Prestel (as at September 1980) can easily be:

(a) Call rates, 3p × 7 days   0.21p per week

(b) Computer access charges, 12p × 7 days   0.84p per week

(c) IP frame charges, say 24p × 7 days   1.68p per week

          Weekly charges   £2.73p + appropriate VAT

Total information-retrieval charges   £141.96p + appropriate VAT p.a.
Increase in TV rental charges,
£12 × 12 months   £144.00p + appropriate VAT p.a.

Thus, for the non-business justified videotex set, the increase to household budget costs could be £285 p.a. (+ VAT). This is excessive and will inhibit growth in this area of the market unless:

(a) Computer access charges are drastically reduced in 'off-peak' hours.

(b) IP frame charges also reduce drastically at all times to residential users.

(c) Rental charges for videotex TV sets reduce in real terms.

To encourage the growth of sets, and usage of Viewtel 202, there will be no frame charges for our national and regional information service. Home entertainments are intended to be self-financing. Advertisers and sponsors will provide all other revenue. If users come to use our service exclusively, or ours plus other non-charging IPs, then they could save some £90 per annum.

If the Post Office would reduce computer access charges outside peak hours to a rate equal to the local-call rate, a further £36 per annum could be saved. Concessions then from set suppliers could mean that the total annual cost could be halved at least from £285 per annum. Regardless of whether charges for the private householder are reduced or not, we are confident that we can provide a wide base of information which will appeal to the remaining business-orientated sector.

2. *Home entertainments.* This will, in time, be operated as a profit centre, with a low IP charge.

3. *Internationally relevent information.* This would be a developing position, with increasing information over a period of time. Already we do have all National Exhibition Centre show and show organizer details. Travel, holidays, business and financial information, and certain types of news information are all probable areas of development.

4. *Specialist (business) applications.* This is a new area so far as we are concerned, but it has already become very clear that this will be an area of major growth over the next 5–10 years. We intend to be fully involved, within the availability of Prestel frames.

To move towards the achievement of development in all these areas, Viewtel now employs three full-time journalists, two full-time marketing/sales and support staff, plus the equivalent of five keyboarding staff. There is also a substantial amount of senior (board) level time spent on guiding and directing Viewtel. The Editor's target is to achieve 300 news-story updates per day, mostly of one frame, but occasionally more. One-frame stories involve 60–70 words. I wonder how many newspapers have a 'news count' anything like 300 live stories per day?

The unique facility of Prestel of being able to interrogate the computer on the number of accesses to any page has already been invaluable in assessing the response to different material put on to our information service. This facility has shown quite clearly that a higher news count gives greater usage of Viewtel. The news can be instant, and, with effort, can compete with any service put forward by teletext, except that their services are totally free of charge. Instant news is ideal for the medium, as well as directory or reference information. However, in-depth, high-word count material is not—hence high-quality editorial précis skills are needed, with the newspaper adapting to re-emphasize its detailed in-depth coverage of events. We do, and will continue to, cross-refer to our newspapers for further, more comprehensive information. At this stage we provide purely factual news coverage and do not offer any comment. This, again, is based upon our objective of as broad an appeal as possible. The news service is a natural extension of the news material already coming into our offices. No reporting staff, as such, is envisaged; only a greater use of the material already available. Editorial skills of selection, re-writing, and presentation can easily be adapted to the new medium. We do not envisage using our editorial staff for inputting—only for re-writing.

We believe that we are consciously and constructively using new technology to enhance our role as information gatherers and disseminators, rather than just as newspaper publishers. However, this broad-based news service will need to be funded, and hence our concept of an electronic newspaper comes in; with editorial providing the appeal and the vehicle, and the advertisers financing the whole service.

Advertising will be broken down into four main categories, equivalent in newspaper terms to:

Directory advertising

Display advertising
Classified advertising
Sponsorship.

These, again, will be broken down into national and regional sectors. Advertisement features/profiles and page sponsorship will also be developed, and cross-references to, and from, other relevant data bases.

Directory advertising will cover, for example: What's on or entertainments; Car dealers; Estate agents; and, in time, all business reference and Yellow Pages-type advertising. These provide simple reference information about suppliers and services in particular sectors, and usually take up half a videotex frame per company. From these half-frames, cross-references follow to more detailed information on complete frames or information trees. This type of development could show a substantial growth area for newspapers. The necessary price of advertising in this media is so low in directory type information—say £25 per half-frame with no updates per year—that many new advertisers would be attracted to it. The equivalent space for one day per week for a full year in the *Birmingham Evening Mail* would cost at least £350, as compared to £25 on our videotex service at present. As a result, less than 30 of the 700 restaurants in the Birmingham area advertise regularly in the *Evening Mail*. Nationally, directory-type advertising shows substantial opportunities, for example, in dealerships and listings of retail outlets and suppliers for particular goods.

Display advertising offers our greatest short-term opportunity—Adflashes— from editorial pages, or from directories, to provide attention catchers. There will also be routing from our advertising directory and also the Prestel directory combined with cross-references in and out of other, appropriate, information data bases. Adflashes already give 4–5 times the page-usage figures of non-referred adverts. Display advertising offers us a unique opportunity to obtain advertising for national as well as regional use. We know we will have a number of totally new advertisers in this area. We can also create 'market places' for particular sectors, for example, retail stores, new housing, and job recruitment. This type of database control and simplification is critical to the growth of usage of videotex information. Regionally, we will complement our major display and classified advertisers newspaper space with a certain amount of 'free' Viewtel space. They will then be encouraged to expand the free space and be charged for the extra. Retail store campaigns will start in the first half of 1981 on this basis.

Classified advertising will occur in a relatively small way at first, on a solus sale basis, and through directories. When our classified newspaper ads are re-formatted to conform in presentation to videotex routing, it will be possible to cross-sell into videotex. and bulk update daily without any re-keyboarding. We are not yet convinced that this type of advertising is desirable or suitable for all categories, but none the less, we must be sure that we can ward off any other entrepreneurs who attack our newspapers through this new medium. Market research will be aimed to establish public and business reaction to classified ads on this medium during 1980/81.

There is one area of major concern for newspapers for the 5–10 year time span. With sufficient sets in the market place—say 15–20 per cent of households—the use of credit cards for mail-order purchases of branded goods in particular, through Prestel or other videotex systems, could seriously affect retail-store advertising. With better graphics and even pictures on cable videotex systems, there could be a major element of newspaper revenue at risk.

Sponsorship. This could become a major revenue area. It would provide the opportunity for companies to sponsor pages of information, with their company associated with those pages.

Viewtel 202 then represents a modified version of the newspaper package—a broad-based news information service supplemented and complemented by extensive advertising and entertainment services. Our information will have to be as relevant and as necessary to the users as we can make it; and at minimum cost. In addition, the future offers major opportunities in business markets, closed user groups, international markets, and in home entertainments. Also, and perhaps more important of all, there is the prospect of running our own commercial videotex service for the West Midlands covering some 2 million households, major business, and industrial areas, and key retail centres. This would be capable of operating 1 000 000 frames plus over 200 telephone ports. Videotex will undoubtedly grow in the UK. We suspect this growth will be faster than for Prestel itself, and largely outside it. Beware the publisher who does not get involved now to ensure a better future for himself. Finally, since videotex is primarily a marketing project, we will be carrying out a major market research project to ensure that we know and understand business and residential users needs and reactions to videotex.

8. Not so much a tele-press

Keith Niblett

Passing decades have seen the newspaper respond with vigour to the unsettling claims of potential invaders. Radio bulletins, cinema newsreels, and television news have all swept into the newspaper's territory threatening revolution. They have all gained ground but a counter-revolution has always developed from within the newspaper itself. It has reviewed its ideas, its appearance, and its presentation to meet the changing needs of a demanding population.

Now Prestel is poised to enter millions of homes and offices with its immediate information service at a fingertip touch. So how does the newspaper respond to the latest innovation? Switch off the presses and go home defeated?

The Prestel potential was recognized in the mid-1970s by Eastern Counties Newspapers (ECN), a regional newspaper group, based in Norwich. The group decided that it could ally itself to this new force in communications, and that Prestel had much wider implications for the communications world than simply as a direct rival to the newspaper.

It was a glossy, inventive, up-to-the-second reference library, something packaged to the size of a television set for the convenience of every front room and simple to operate in spite of its technically complex basis. It was a development in which the group wanted to be involved and it offered fulfilment of a long-held desire to enter the field of electronic communication.

But the ground was not completely secure. Perhaps, even now, the foundations are still unsure. Certainly, ECN had to weigh the elements very cautiously before stepping in with investment. Was Prestel ready to lead the world into a new technological era? Might it be nothing more than just a passing whim, a stage of technology which grips the mind with its potential yet is surpassed just as quickly by even more amazing refinements? Would Prestel take off with such power that newspapers could be extinct within a matter of decades . . . or would it flop so disastrously that a few months would see it viewed only in the British Museum?

Somewhere between the two rested the answer. ECN decided that Prestel had a future and that the group wanted to be part of it. ECN's first step into Prestel was made in 1975 and was seen initially as a direct extension of the newspaper publishing activities which produce a morning, evening, and series of weekly papers. Experience quickly proved, however, that Prestel is not an electronic newspaper but something much wider. The medium was so different that it could best be developed in a semi-detached form. Prestel therefore became divorced from the other operations at ECN, although the involvement has only been possible in economic terms because of the association with the newspaper publishing side of the business.

The second, rather more remarkable, factor to emerge was that no one had put a finger exactly on what Prestel was for! The public were being launched into a world of keypads and buttons without knowing precisely what they had to gain—or lose. Even those involved in providing the information to Prestel were uncertain. But Prestel users would want to know that what they were about to receive was something special; something worthy of a not inconsiderable cost. Prestel is not a one-armed bandit but it is not cheap—and that may prove its biggest hurdle.

The initial launch of Prestel was inevitably aimed more towards the businessman. But ECN was looking ahead to the time when Prestel was available to and used by the general public. Under the name of Eastel, the company became an information provider to Prestel offering a database of general information and entertainment, limited news and public service material, and general advertising. Classified advertising fitted naturally into the Prestel slot and ECN began preparations for a comprehensive database in all the major classifications.

News content is more difficult to handle. The average newspaper reader likes to scan a page, to browse through a story and to follow in detail certain items of particular interest or appeal. Prestel cannot go far beyond the headlines and bald facts of the news with little room to elaborate or decorate.

So Eastel has provided a range of information aimed not just at the people of Norfolk but at potential Prestel users all over the country. A prospective visitor to Norfolk will want to know the times and modes of travel, where he can stay, where he should visit, where he can eat and what is on the menu. The answers are there at a glance and ECN has been praised for its presentation. So the investment has been made but the stakes are high and healthy returns will depend on further development of the system, the rate of growth, and the audience response.

The information providers inputting into the system do so for a variety of reasons. There are those who want to market their own products; those distributing information from their own business; those buying and collecting information to reprocess on to Prestel; and those offering expertise and database building services to other companies wishing to join Prestel. Eastel is one of the few information providers which actually fulfil all four of those functions.

ECN has defined its role as being: to market and defend its own products; to use the information gathered within the organization; to act as a brokerage for advertising customers; and, using its expertise gained since 1975, to act as an 'umbrella' to other organizations.

Companies already using Eastel's services include television retailing chains, a furniture group, and an American media corporation. Interest has also been great from other publishing groups both in Britain and across the world. The Norwich base of Eastel has been 'open house' for a constant flow of visitors.

As an information provider, Eastel has had to explore the ways it can best present Prestel to the public, how to make the most impact on the screen and where the real future of the medium lies. The guidelines on which it works realize

115

that graphics are limited, that information must be concise and easily read, and must be found without recourse to flitting through a series of frames of pages. But the most important area of exploration has been to find and then exploit the areas where Prestel is decidedly superior to the competitors around it.

There are some areas where Prestel steps out of the crowd. It is immediate. News and any other information can be updated much more quickly than in, say, a newspaper. It also has the benefit that the amount of information held in the system is virtually unlimited. It can therefore offer access to a quantity of data which in other forms would be quite unmanageable.

The most tempting attraction of Prestel is the facility which allows the viewer to answer back, instantly. This creates spontaneity and provokes reaction from the customer. It is only necessary to watch children playing games on the database to guess how positively the next generation may react to electronic media. There have already been forecasts that by the end of the century—perhaps sooner—Britain will be a nation of armchair shoppers. Some shops have reacted quickly and already list their stock on Prestel. Can it be long before the Prestel user simply keys his credit-card number into the system, records his order, and sits back to await delivery? It sounds like the housewife's dream; to order a month's supply of groceries from the supermarket and then put her feet up until the goods arrive. The marketing men will surely not miss such an opportunity.

It is an area, too, where the supposed competitors, newspaper and Prestel, can happily operate alongside each other. A department store holding a sale could place an advertisement in the newspaper giving notice of the event. It could, at the same time, refer to Prestel pages where stock items for sale would perhaps be updated by the hour. With the interactive order frame facility, there would be no need to queue outside the department store. It is a situation where ECN would be able to offer a complete advertising package to the department store. It means that the traditional newspaper browsers are alerted to an event, with the detail being provided by the selective, fast, updated medium; the new and old technologies working together as a compatible force.

It already appears that videotex, in any of its forms, will not greatly affect the newspaper editorial services. But Prestel is steadily emerging as a suitable medium for classified advertising. It follows that, both as a defensive and money-earning measure, Eastel should offer a comprehensive database in all the major classifications. If it does not, other companies will.

Prestel should be self-financing and ECN is confident that initial investment can be met by contributions from the new marketing opportunities presented by a new medium. The most important among these was the *Prestel Users' Guide and Directory* which was launched in 1978 and was revamped as *The Prestel User* two years later. It was clear from the start that any service offering tens of thousands of pages of information should also provide a comprehensive guide for the customer, in the same way that the telephone has its directory and Yellow Pages.

Backed by the Post Office, ECN were first on the market with a complete index of information providers and what they had to offer on Prestel. The directory was well received and became the forerunner of similar publications. It incorporates a magazine section, designed to educate, entertain, and interest Prestel users, particularly when more of them are drawn from the general public. Another new outlet was the group's consultancy, known as Eastel Services. The consultancy is able to provide a full service for sub-information providers. The steady growth of Eastel has caused the viewdata staff to increase. Recruitment has generally been internal, with specialist knowledge being acquired over a period of time rather than bought direct from a market where the supply is very limited.

By the beginning of 1980 the department had grown significantly. Beneath the viewdata manager the department is divided into three sections. The first is viewdata publications which, on the sales administration side, employs three people. Another three, an editor, an artist, and a works make-up man, are drawn from the internal resources of the company and work for about 30 per cent of their time on the project. Secondly, the Eastel database employs an administrator, a database designer, and a liaison officer, who deals with both internal and external workflow problems. Thirdly, Eastel Services has a salesman/consultant who uses the database team for his throughput of work.

Early in 1980, ECN had six on-line inputting terminals with plans to invest in an off-line editing/bulk updating system. Under the guidance of a supervisor, the staff comprised two full-time and four part-time inputters working round the clock. A further capital investment of more than £100 000 is expected in the future. Based on micro-computers and with additional software, much of it designed internally, the outlay will provide four off-line editing terminals at Norwich and another at the ECN office in Ipswich.

The equipment will also allow classified newspaper advertisements to be drawn from computer store and reprocessed into a database tree ready for Prestel acceptance. If Prestel lives up to the expectation of popular forecasts, the company should recoup its investment within five years.

The service provided to Prestel by ECN will reflect the standards of its traditional role as a newspaper company. That means a complete update of local advertising, news, and information at least once a day.

Absorbing Prestel into an existing operation has presented its problems. Nearly every department within the company has seen its workload increase, from front counter clerks, salesmen, administration, and accounts through to the printing union cardholders who have formed the inputting department. The Prestel system was built by engineers and not by behavioural designers—and the display screen limitations are obvious.

Prestel is a marketing medium. Information on Prestel has to be marketed. To do that well means good presentation and that is where the role of database building becomes vital. Information must be easily found through good indexing. It must give all the relevant details in a precise form with a clear and attractive presentation. The user, often unfamiliar with the system, must be provided with an easy escape route if he loses his way deep in the heart of the service. At least two options should be given on what to do next. Graphics should be used as an integral part of the information—not merely as decorations or space fillers.

Much has been learned about how to do this but there is a long way to go. There are no old masters on Prestel. ECN has helped promote and develop the system without detriment to its range of newspapers; in fact a harmony has been created between the two.

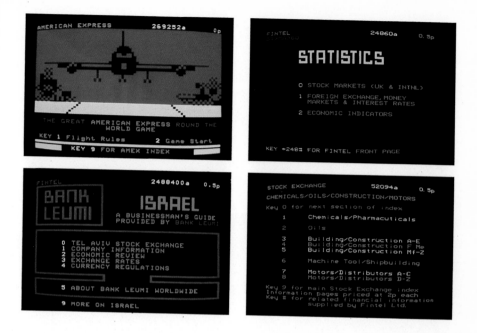

1. Viewdata systems work by showing the user a series of indexes, from which a choice has to be made. There are many ways of showing an index on the screen—indeed, the computer itself recognizes no such category, since any page can have as many or as few onward choices, up to ten, as the creator of the page wishes.

The possibilities range from the bare no-nonsense listings, as in the London Stock Exchange industrial sector classification and the Fintel statistics index, through the mixed index and graphics of the Bank Leumi page, to the American Express aeroplane, which is both the start of a game, hence fun to look at, and the opening index on how to play the game.

Given the heavy overhead of indexes on any viewdata system, it is likely that more and more attention will be paid to mixing indexes and information, or combining indexes and pictures, so that the pure index page will tend to wither away.

Network Data Ltd   22423a   0.0p
OFFICES : CITY EC1,2,3,4   AT FEB 8
1106 FOUNTAIN HOUSE 130 FENCHURCH ST
     REFURBISHED TO A HIGH STANDARD
     10TH FLOOR 7 YR LEASE £20000PAX
     CONRAD RITBLAT   01-623 9116
1150 14 DEVONSHIRE SQ EC2 -B/C GROUND
     FLR RECENTLY REFURBISHED-ALL
     AMENITIES £13225PAX LEASE 87RR82
     CONRAD RITBLAT   01-623 9116
1352 SHORT TERM OFFICES UNTIL END
     1979 RENT £5950
     ROBERT CUTTS   01-236 4606
1450 LUPGATE HILL 3RD & 4TH FLOORS
     LEASE B/A RENT £8500PAEX
     MICHAEL LAURIE   01-493 7050
     HILLIER PARKER   01-629 7666
KEY ## TO RETURN ,KEY 0 FOR NEXT FRAME

FINTEL   53134640a
FITCH LOVELL
         RECENT DEVELOPMENTS
9/5 Key Markets, the food retailing
    subsidiary of Fitch Lovell is to
    pull out of discount food stores.
28/7 2nd half pre-tax profits were down
     to £3.47m leaving the total for the
     yr to Apr. 78 down £1.46m to £6.23m
5/9 The chairman has said that results
    for the current year should be not-
    iceably ahead of those for last yr.
29/9 The AGM was told that the manufact-
     uring and poultry divisions had
     made a good start to the year.

KEY 1 FOR MORE ON THIS COMPANY
    9 FOR MORE ABOUT FOOD RETAILING

BARIC   (C)1980 35010000a   0p
Pan Am   STANDBY   747
                  Updated   747
               12/2/80 1245Hrs   747
Wed 13th Feb 1980
Flight Destination Time   Chance Fare
PA101 New York   1100   Good £85.50
PA101 Detroit   1100   Good 96.00
PA101 Washington   1100   Nil   91.00
PA001 Miami   1115   Good 105.50
PA125 San Francisco 1210   Good 115.50
PA125 Seattle   1210   Nil 111.00
PA107 Washington   1325   Good 91.00
PA001 New York   1400   Good 85.00
PA001 Houston   1400   Good 93.00
PA121 Los Angeles   1440   Good 115.50
Standby seats availability subject to
weather conditions.
Standby seats are on sale from 0600 hrs
PanAm Air Terminal,
Henley Place, Victoria, London SW1
Full check-in Facilities & Coach
Services Available   Key 0 PanAm Index

Meteorological Office   209410a
OVERSEAS   Weather   1200 GMT.
Europe(N&W)   12/2/80
LOCATION   °C   °F   WEATHER
Amsterdam   7   45   fog
Biarritz   11   52   sunny
Bordeaux   9   48   partly cloudy
Boulogne   8   46   partly cloudy
Brussels   6   43   mist
Geneva   7   45   sunny
Locarno   10   50   sunny
Luxembourg   5   41   mist
Lyons   3   37   mist
Marseilles   13   55   partly cloudy
Nice   15   59   sunny
Paris   7   45   mist
Perpignan   14   57   partly cloudy
Strasbourg   7   45   partly cloudy
Zurich   7   45   sunny
Europe: 0 S&W   1 S&E   2 N&E

2. The viewdata page most naturally lends itself to lists of various types, and a large proportion of the material on Prestel, for example, is of this type. This tends to emphasize the reference function of viewdata, as a place where all sorts of information can be easily 'looked up'.

Examples here range from weather reports for various cities provided by the Meteorological Office, to lists of office blocks available for occupation in the City of London, availability of standby seats on PanAm flights to America from London tomorrow, and abstracts of business news stories that have appeared in the *Financial Times* newspaper.

These pages can change with different frequencies—the PanAm and Met Office pages change daily, by definition, whereas the *Financial Times* abstracts page, once created, remains unchanged until it is deleted as being of no further interest. But the style of the page remains essentially unchanged, as does the nature of the material on it and even its placing on the page. There is a skill, not always evident, in presenting lists so as to be immediately informative to the user.

```
FINTEL                    331660a
         GOLD BULLION PRICE            10p

              $ per fine oz

                    FEB        FEB
                    12th       11th

  OPENING          706.50     718.50

  MORNING
  FIXING           703.75     714.50

  AFTERNOON
  FIXING           694.50     710.50

  CLOSE            697.50     712.50

KEY 0 DAILY GOLD PRICE FROM JAN '79
    8 MORE ON COMMODITIES
    9 UPDATE TIMES INDEX 9 UK STATS
```

```
                              5712140a      2p
G.S.S.  Government Statistical Service
INDEX OF RETAIL PRICES
All items
January 1974=100

      1979  1978  1977  1976  1975  1974
Jan  207.2 189.5 172.4 147.9 119.9 100.0
Feb  208.9 190.6 174.1 149.8 121.9 101.7
Mar  210.6 191.8 175.8 150.6 124.3 102.6
Apr  214.2 194.6 180.3 153.5 129.1 106.1
May  215.9 195.7 181.7 155.2 134.5 107.6
Jun  219.6 197.2 183.6 156.0 137.1 108.7
Jul  229.1 198.1 183.8 156.3 138.3 109.7
Aug  230.9 199.4 184.7 158.3 139.3 109.8
Sep  233.2 200.2 185.7 160.6 140.5 111.0
Oct  235.6 201.1 186.5 163.5 142.5 113.2
Nov  237.7 202.5 187.4 165.8 144.2 115.2
Dec  239.4 204.2 188.4 168.0 146.0 116.9

Key: 1 Index of Retail Prices (all
       items except seasonal food)
     0 Prices index page 0p
```

```
SARIC              (C)1979 350451294a   10p
JORDANS TOP 500 UK UNQUOTED COMPANIES
MORRIS & DAVID JONES LTD.
(Sub. RCA Corp. - USA)
87 Gt. North Rd,Hatfield,Herts.AL9 5EG
Ch/ChExi J.C. Page
Distribution & food processing

£000,000           12/78   12/77   12/76
Sales              67.9    76.0    82.8
Exports            0.1     0.1     NIL
Pretax Profits     1.8     4.0     0.88
Fixed Assets       8.4     7.5     8.5
Current Assets     18.0    16.6    19.0
Current Liabils    12.9    12.3    18.2
Net Assets         13.5    11.8    9.2

UK Wage Bill       4.5     4.3     4.3
No.UK Employees    1712    1898    2019

Key 9 For Survey on Foreign Owned Cos.
For Jordans Index <0>  Supermarkets <1>
For Canned & Frozen Foods Companies <2>
```

```
STOCK EXCHANGE                 520019a   2.0p
CHEMICALS/PHARMACEUTICALS
                      CLOSE
ALD. COLLOIDS         83     2
BEECHAM               603    5
BOC INT               69     ½ 9 ½
BRENT CHEM.           204
CARLESS C             29     8
CRODA INT             56
FISONS                297    9
GLAXO                 481    3 1
HICKSON W             198
ICI                   353    4
LAPORTE               107
LEIGH INT.            122    1 19
RECKITT C             457    2 49 52 1
RENTOKIL              70½
YORK. CHEM.           85

FRAME UPDATED ON 25/01/79 AT 15.44
KEY 0 FOR NEXT PAGE, 9 FOR MAIN INDEX
```

3. Because viewdata lends itself to lists, much of the information, at least for the business user, tends to be in the form of statistics. Colour can be used to make the display of the statistics more intelligible, and to make up for the limitations of the viewdata page. Even so, there is a question about how many statistics to pack on to one page, and what price to charge for the resulting quantity of figures.

Here there is a tightly packed page from the Central Statistical Office (a government organization) showing the movement of retail prices; company financial data from Jordans; a page of Stock Exchange prices; and the London daily gold price provided by Fintel.

Here again the different rate of change of the information is evident. The gold price is updated four times a day; the share prices (at present) three times a day; the company data as required; and the retail price data is simply added to at regular intervals.

4. Pure text—that is, filling up the screen with words from top left to bottom right—is not often used, although for certain purposes, in business use for example, pages like the one provided by the *Economist* weekly magazine can be very functional. Otherwise, pure text can be hard on the eye: viewdata is probably not a reading medium in the sense that a book or journal is.

More often, the text tends to be modified in a variety of ways, to alleviate its effect. The Extel Sports football page has adopted a schematic layout so as to sectionalize the reading matter: the Florida hotels for American Express takes on the look of a holiday brochure: and the weather forecast combines a map and text.

Where pure text is used, it is important that each page be self-contained, and not run on to the next in the middle of a sentence, and that colours be used in a restrained way. As with indexes, it is likely that pure text will always tend to give way to a new style of viewdata page that combines, as do several of the illustrations, the various viewdata elements.

5. Company logos are at once an attractive and a limiting feature of viewdata. Corporations with a house style which they like to see repeated in all public contexts may or may not find that the graphic facilities on viewdata can cope with their particular requirement. It may be that pressure for upgrading Prestel and allied systems in respect of their pictorial capabilities will most obviously come from easily made comparisons between designs, such as company logos, that exist prominently elsewhere and need to be reproduced on viewdata as accurately as possible.

The examples show attempted logos for W.H.Smith, a newspaper and book retailing and wholesaling chain; for Cable and Wireless, the UK-based international telecommunications operator; and for two banks, Barclays and Société Generale de Banque. They are more or less successful, according to taste. But the difficulty of drawing curves on Prestel-style systems is very apparent. There is also some argument about the wisdom of putting the company logo on the front page of the viewdata 'magazine' anyway, since it takes a 'printed' view of the front page as opposed to a 'dynamic' view that recognizes the user activity that the page represents.

6. Graphics should be more than pure decoration, although there is a strong temptation to put them in just for show. The fact is that in most circumstances there is no such thing as free use of viewdata— even if the pages are free, there is usually a charge for using the telephone, or for accessing the computer, or both. So users will probably not pay just to look at second-rate pretty pictures. Purposeful graphics are another matter.

The examples here show a game, in which the Foxes have to catch the Goose, where the graphic is an intrinsic part of the game; a performance graph from Datastream, where the graphic is at the same time the information content; a bar chart from Fintel, where again a different style of graphic presentation is used to convey the information; and a mixed text and graphics page from Eastel (alias Eastern Counties Newspapers, of Norwich) which uses graphics to enhance the textual message.

Used in this way, graphics are not 'stand-alone' visual effects, but are part of the information mechanism of viewdata. As before, it is likely that graphics as a separate element will tend to wither away as they merge into the general art of viewdata page design.

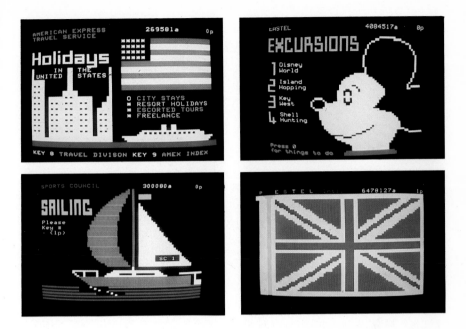

7. The necessary qualifications having been made, graphic effects are among the most attractive features of viewdata, and most readily lend themselves to demonstration and reproduction. As discussed in the main text of the book, various viewdata systems have (or claim) varying capabilities in graphics. Here are four Prestel graphics associated with indexes (again, the mixed-function page). They range from the Union Jack to Mickey Mouse, from holidays in New York to holidays on a boat.

It is probably true that even the business applications of viewdata, or at least the public applications to business data, cannot ignore the graphic aspect, since even businessmen are human. (Private in-house applications may, or may not, be different: to treat viewdata as a pure data processing activity is possibly to waste some of its best features.)

Editors and managers of viewdata information services are gradually learning to drop preconceptions derived from conventional computer systems, newspapers, books, brochures, typewriters, or advertising studios, and think afresh about viewdata presentation.

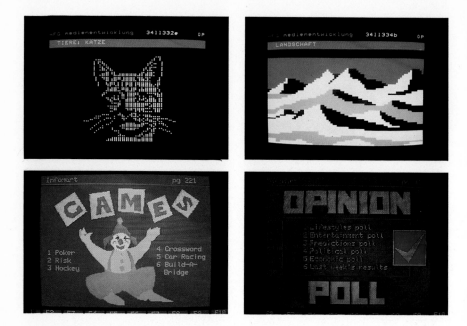

8. Different styles of visual presentation are shown in these two sets of two illustrations, one set taken from the German Bildschirmtext system, the other set taken from the Canadian Telidon system. Bildschirmtext is a modified version of the Prestel system, and these graphics probably push to the limit the possibilities of Prestel-type systems as they are at present (that is, without taking account of possible future enhancements, such as 'Picture Prestel' which allows transmission of real pictures to the Prestel screen).

The Telidon screens show the quite different picture resolution made possible, at a price, by the Canadian design. As with the Japanese Captain system, also discussed in the main text of the book, the philosophy can be said to be quite different. Whereas Prestel, in layman's terms, sees itself at present as a medium for text and figures, with the quality of the graphics being a result of the standards set for text transmission and display, the Canadian and Japanese systems can be said, again in layman's terms, to be fundamentally displays of pictures the size of a TV screen, part of the content of which happens to be letters and figures.

Information medium or picture medium? Or is that a false question, based on a false dichotomy? Here, as elsewhere, viewdata raises basic issues of communication.

9. The economics of closed user groups

Stephen Castell

If you are an owner or manager (actual, or potential) of some specialist body of business information, and feel you would like either to found a new business on, or enhance/diversify an existing one by selling this information electronically, what are the types of questions you might ask of Prestel, and how might the answers compare with those obtained from competitive methods (if any) of electronic information dissemination?

In attempting to give an answer to this question, I specifically exclude comparisons with 'traditional' publication in print media—noting, nevertheless, that, however much the pundits of electronic publishing may enthuse about such new media, one ought not automatically to discard all thought of 'wood-based random acces information products' (i.e., books, journals, etc.)!

Very loosely, I mean by 'specialist information', information 'supplied by specialists, for specialists'. By '*business* information' I mean that which is, in subject, content and presentation, as far as possible away from the 'Irish jokes' database, say, on Prestel (in the sense that Irish jokes have, supposedly, universal appeal—even, I understand, to Irishmen, whereas business information essentially does not).

One might further define that which I am taking great pains *not* to define by reference to the classic story used by psychology professors to illustrate the distinction between 'psychology' and 'psychiatry'. Thus: a beautiful woman enters, and proceeds to walk across, a room full of men; all heads turn to watch her passage (*sic*). All, that is, save two: one of these men is watching the *other* men watching the woman, and the second man is watching the *first* man watching the other men watching the woman. The first man is the *psychologist,* studying normal human behaviour, while the second is the *psychiatrist,* studying abnormal human behaviour.

In a similar way, we could imagine two viewdata business information suppliers seated at their Prestel editing input terminals. One is busy tapping away, entering corporate, travel, econometric data, etc., from an input sheet, while the other is looking intently at the first and occasionally entering details of his (the first's) activities. The first IP is undoubtedly the non-specialist business information supplier—entering information about completely *normal,* and general, business behaviour; the second is, however, most definitely the *specialist* business information supplier—entering information about highly abnormal (or, at least, esoteric) business behaviour.

In general, the scope of the questions which such a specialist business

information provider would ask of Prestel on contemplation of his becoming an IP might then be:
- what are the basic participation costs?
- what is the level of specialist technical expertise required, and its associated costs?
- what is the management and control effort required, and its associated costs?
- what are the 'product' problems and costs: e.g., sourcing, production, quality control, delivery, etc.?
- what are the marketing problems and costs?

Prestel—the basic costs

For the purposes of evaluating and comparing competitive costing structures, I take as an illustrative model a specialist information supplier who wishes to book 1000 frames p.a. on Prestel to mount his specialist database. To attempt to put this in context, it seems to be something of a consensus that 1000 frames of coherent business information can be considered a 'medium'-sized database on Prestel, in the sense that anything less is 'small-to-medium' and anything greater 'medium-to-large'; also, anything between 15 and 30 per cent of such frames are likely to be used for routing, indexing and cross-referencing—for 'structuring' the data—with the remainder purely for 'end data'.

One should note in passing that, while I may happily take this as an illustrative model, the hopeful IP may actually find it currently difficult to obtain 100 frames under his own 3-digit node on Prestel, let alone 1000. If the Post Office is to be believed, practically all of the 250 000-frame capacity of the public-service computers is booked by the existing 150 IPs, with perhaps 400 further interested organizations already on a waiting list, pressing to be accommodated. This situation is, however, expected to be radically eased in 1981 with commissioning of additional computer disc storage.

Based on the published PO tariffs, the new IP's costs would be along the lines shown in Table 9.1. This demonstrates total direct Prestel costs of £14 030 for the first year. If no 3-digit node storage is available, as I have indicated is likely to be the case until well into 1981, the new IP has the alternative of appearing on Prestel within the frames of one or other of the existing 3-digit entry-point IPs who deliberately booked large quantities of frames in the early days specifically to be able later to sub-let this space as an 'umbrella' organization.

Such organizations are at the moment understandably coy about publishing hard tariffs for their umbrella services, but it appears that per frame all-inclusive figures varying wildly between £15 and £50 p.a. have on occasions been quoted. For the purposes of our illustration, it would seem prudent to take the higher figure, thus giving an 'all-in' cost of £50 000 p.a. to the new IP for his 1000 frames taken under a Prestel 'umbrella' organization. For this sum he should of course get from the organization involved a large amount, if not all, of the design, creation, and maintenance of his database included, together with a degree of overall consultant–client 'hand-holding'. This should be borne in mind when comparing the £50 000 with the earlier £14 030 total, and when I come to look at

Table 9.1 Public-service tarriffs for specialist Prestel closed user groups

Assumptions
1. 1-year booking basis (N.B. discounts for longer-term bookings are available).
2. 3-digit node entry-point—'A' Class storage.
3. 1000 frames booked (N.B. 100 is minimum lot available).

Storage	£4 per frame p.a.
Service charge	£4000 p.a.
Additional CUG Service charge	£4000 p.a.
Editing terminal	£1630 initial payment + £100 per quarter rental.
Connect-time computer usage	Currently free of tariff charges to the IP when using the editing software, but IP is charged as any other user when inspecting his database to check it after editing, or when doing demonstrations.

On 1000-frame booking, first-year costs are then:
£4000 + £4000 + £4000 + £1360 + £400 = £14 030 total

(N.B. 1 frame = 24 lines × 40 characters = 960 characters
= approximately 750 'usable' on-screen characters maximum, allowing for space lines, graphics, etc.
Then at £4 per frame, cost of storage per character
= 400p/750 = 0.53p per character p.a.)

the other operational cost areas involved in running the specialist IP's business.

While dealing with Prestel IP tariffs, one must of course take into account that Prestel *user* tariffs will no doubt affect the pricing policy of the IP's own service to his market—as will, indeed, the choice and costs of TV-terminal types on offer by the manufacturers/renters, and general set availability. Leaving aside, however, TV set rental or lease costs to the user, the PO charges the user a peak (and therefore business-hours) rate of 3p per minute connect-time, and generally allows the IP to set a price on each frame of between 0 and 50p (which revenue goes direct to the IP).

Comparative costs of other alternatives

Having dealt with Prestel IP tariffs on the assumption that our specialist information supplier has already decided on Prestel as his electronic publishing medium, let us now go back a stage and ask: is Prestel the only way to fly? The characteristics of the specialist information supplier's business, and the range of options of 'positioning' his service in his market, made this a valid question to ask.

There are appearing a number of videotex-type alternatives to the British Post Office's pioneering Prestel viewdata service. There is at least one private viewdata system being offered in the UK to potential information suppliers hungry for frames, and most user terminals in the field in the UK have more than one telephone autodialler 'channel' so that alternative non-Prestel viewdata services may be accessed through the same hardware. The PO itself has sold its viewdata

121

software to several other countries, notably West Germany, whose 'Bildschirmtext' service is now at an advanced pilot stage. Non-PO-viewdata videotex systems (e.g., the French Teletel and the Canadian Telidon) are also developing spontaneously within several other countries' PTTs. In the US, Insac, the UK's National Enterprise Board 'software marketing' subsidiary, has signed a joint agreement for exploitation of Prestel software with GTE (General Telephone and Electronics).

In the UK, the PO has asked for offers to lease frames on a new 'closed user services' computer, distinct from Prestel. It will be sited in London and have initially a 100 000-frame and 80-communication port capacity.

Returning to Prestel, but still away from Prestel UK, one of the most interesting departures, particularly for specialist business information suppliers, has been Prestel International. Although this is PO initiated and controlled, it is being developed under conditions and management 'rules' quite different from those which have evolved for Prestel UK.

All of these viewdata-based options may well be considered by our specialist business information supplier. Almost without exception, none of them, however, has yet achieved a public-service status, and it would be idle speculation at this stage to attempt to present anticipated tariffs.

A more practical example of a system alternative is afforded by the regular computer-timesharing bureaux networks. These have traditionally been providers of remote-access processing power, to be used by trained computer specialists or intermediaries and, compared with the general run of viewdata usage, with high access costs. For the specialist information supplier, however, this type of system could well be considered as a valid alternative. Indeed, the presence of processing power (compared with viewdata's 'data look-up' facility only), frequently commanded by a high-level computer language, may well be a positive benefit for the type of information service he is intending to supply. Nor should one underestimate the 'communications reach' of these established timesharing bureaux—both of the companies chosen for illustration, Rapidata and ADP, operate a network of more than a dozen DEC-10 mainframe processors (as well as very many more mini-computer communications 'front-enders'), and could each have a capacity of 5–10 000 terminals world-wide.

(One should perhaps not overrate 'communications reach', however—it is after all the message, not the medium, which is ultimately important. I am reminded of the story of the explorer, lost in the densest of jungle, who, practically on his last legs, hears the distant thump of a tom-tom. He follows the sound for five days and, with his stamina failing fast, finally comes upon a clearing. In the middle sits a huge African chief, in his left hand a bright shining Driver and in his right an equally sparkling No. 3 Wood. With these he is pounding the leather out of a large drum. The explorer crawls across the clearing and just has the strength left to scratch the chief on the ankle, croaking '. . . Heard the drum from five days out. . . . Followed the sound here . . .', before expiring. The chief, ignoring the unfortunate ex-explorer, looks admiringly at his Driver and then his No. 3 Wood, and says simply 'Five days, huh . . .? Man, am I getting some length with these new clubs!' Is there occasionally something of the chief's insensitive approach in the attitude of Prestel partners to their new viewdata 'club'?)

Table 9.2 shows some comparative storage, processing, and usage costs for Rapidata and ADP. I think that the costing structures they (and no doubt other comparable network-facility operators) are prepared to offer show at least that they are worth evaluating as a realistic system alternative for the putative specialist business information supplier.

Table 9.2 Comparative costs of other system alternatives

Examples: *Two on-line timesharing bureaux networks currently becoming heavily involved with database applications. Each has upwards of 1 dozen networked DEC-10 mainframe processors, giving a world-wide on-line terminal capacity of possibly 5000–10 000 terminals.*

	Rapidata	*ADP*
Each user number	£5 p.m.	£5 p.m.
CPU usage	3½p per 'Central processor unit'	14p per 'Processor resource unit
Connect time: 120 ch.p.s.	£6.50 p.h.	£8 p.h.
Maximum on-line storage	30p per 1000 char.	2p per 1920 char.
charge rate	Unit p.m.	Storage unit p.d.
	= 0.36p per char p.a.	= 0.375p per char p.a.
Probe storage rates (i)	0.06p per char p.a.	——
Up tp 75% discounts (ii)	——	0.09p per char p.a.

Notes:
  (i)   Special storage structure specifically designed to go with Rapidata's Probe database manipulation and retrieval software.
  (ii) ADP offers discounts on storage of up to 75 per cent for large long-term commitments.
  (iii) Both have off-line storage rates for non-volatile data of the order of 0.01p per character p.a.
  (iv) Both are interested in discussing new databases with any information owner (numeric or statistical preferred, but bibliographic also a possibility).
  (v)   Both are prepared to offer free set-up, free storage, and free maintenance to any information supplier who will provide database and help to market it.
  (vi) Both quote typical user processing costs of £30–£100 p.m. guideline.
  (vii) At least one is looking at viewdata software packages on DEC machines.

Some possible characteristics of the specialist information supplier's business

Whether our purveyor of esoteric business information chooses Prestel, an alternative videotex service, regular timesharing bureaux, or some other electronic publishing medium, or, indeed, a combination of these, he will need to determine and define the characteristics and 'style' of his business, have a good prior estimate of the overall costs of running it, and try to make a reasoned attempt at defining a pricing policy for the services it will offer which will make it highly acceptable to its market. All these aspects are, of course, related. The salient points of an analysis along these lines are presented in Table 9.3, covering

Table 9.3   Some possible characteristics of the specialist IP's business

---

(a) Business style
- close contacts with the market-place: professional needs and level of sophistication of users;
- good familiarity with how business is done in the profession or industry of his market-place;
- presents as a 'top-quality business intelligence service';
- importance of 'quality control' of his database product;
- provision of *hardware* as well as information service?

(b) Revenue-earning opportunities
- subscription base;
- direct information retailing electronically (by Prestel or other);
- information wholesaling by joint venture;
- specialized information sourcing and consultancy;
- managing others' information: 'every customer is a potential supplier'.

(c) A first-year business plan
1. Costs budget (£)

Information abstraction and database maintenance	30 000
Prestel or other on-line facility costs, say	30 000
Management + staff + associated overheads (4 people)	60 000
Office and establishment expenses	20 000
Marketing and advertising	50 000
Miscellaneous overheads	10 000
	£200 000

2. Income targets (£)

Subscriptions: 200 at £1000 p.a.	200 000
(or 2000 at £100 p.a., or a mix, or 'happy medium')	
Usage revenues: say 200 users at £15 p.m. usage	36 000
	£236 000

Other revenue earning areas are largely unquantifiable until specific opportunities are researched and pursued:
- information wholesaling
information sourcing/consultancy
information management/'umbrella', etc.

'Business style', 'Revenue-earning opportunities', and 'A first-year business plan'. The only areas from this overall table I would pick out for emphasis are these:
- information *sourcing,* rather than direct database supply, may well be the most powerful 'product' the specialist information supplier has to offer: several studies tend to agree that the busy businessman and professional actually prefer fast and reliable *access* to top-quality *sourcing* of primary information, rather than an electronically-available database of the primary information itself 'on his desk at the touch of a button . . .' (this seems to support a perhaps paradoxical view therefore that, as far as specialist

business information goes, even Prestel may come to be used more by the specialist 'mid-user' intermediary rather than the untrained end-user professional or businessman);

-   the possibility of managing ('laundering'?) other people's small, specialist areas of business information should not be lightly discounted. The description of 'every customer being a potential supplier' is perhaps true for the information 'product' as for no other product in the history of marketing. The simplicity of electronic publishing, under the co-ordination of our specialist business information purveyor, offers commercial opportunities to many small organizations, for whom traditional media are too costly and cumbersome—and these organizations are just as likely to be among the most enthusiastic users of his services;

-   in the first-year costs budget, the sum estimated earlier for a Prestel 'umbrella' solution to the problem of database establishment, maintenance and management, £50 000, would probably be spread over the first three headings of account shown in Table 9.3; overall, the costs budget should be about the same.

Example: Infolex, a specialist business information supplier on Prestel

As a concluding illustration of the type of business I have in mind in this article, I should like to mention briefly as an example, Infolex Services Limited (a company in which I have to declare an interest as co-founder). Infolex is one of the first IPs, contracted with the PO since October 1977, and was specifically formed earlier to provide computerized legal information retrieval and other communications services for practising UK lawyers, which it will do on a closed user group subscription basis to a total market of perhaps 15 000 British legal researchers of the type envisaged.

The principal occupant of Infolex's 1000 frames on the Prestel database is CLARUS, a Case Law Report Updating Service, now giving approximately 2000 up-to-date references to legal precedents and judgements arising from British court decisions as reported in the well-known standard and respected law-reporting journals published in England (e.g., *All England Law Reports, Weekly Law Reports. Times Law Reports*). Abstraction began in May 1977 and now covers five of the major reporting sources.

The case references are presented on the Prestel frames in a form very familiar to UK lawyers, and are easily retrieved by use of a unique 200-term Infolex index structured over the viewdata frames with a businesslike 'no frills' presentational philosophy in order to meet the objective of a practical, cheap, and fast updating service for the busy professional legal researcher. Infolex has also recently begun development on Prestel of its second main database STALUS, a Statute Law Updating Service, with initial coverage of an important piece of UK legislation, The Consumer Credit Act, 1974.

The usefulness of the Infolex service depends on the scope of material covered, the depth of reporting (and case summaries, to supplement bald references, for example, are a planned addition to CLARUS), and the 'finesse' of the index. It is proving of practical benefit to lawyers, at an economic price—borne out by the spread of legal research interests represented by the initial Infolex subscribers, which include private firms of solicitors, corporate lawyers, patent agents, and legal educationalists. It is also of interest to note that Infolex clients tend simply to see Prestel as the cheapest, most cost-effective medium for obtaining the particular information, rather than expressing strong interest in the overall Prestel database as such.

To put the development of Infolex in context, it may be noted that it is the first such commercially available computer-assisted legal information retrieval system to be launched in the UK. Full-text computer-based legal retrieval systems such as Lexis and Westlaw, together with a newcomer, Eurolex, based on the Status retrieval software developed in Britain several years ago, are also planned to be made available by other operators in the UK in the early 1980s. These are likely, because of their immense full-text retrieval power, to be many times more costly to subscribe to and use than the Infolex service.

In addition, it is worth noting that the seminal report, 'A national law library—the way ahead', published in February 1978 by the innovatory UK Society for Computers and Law, has been followed up by the establishment of a company, The National Law Library Ltd., which is currently considering alternative strategies and features of an optimum computer-based service for the UK legal profession.

The future

As videotex and other electronic publishing or public computer-based information systems proliferate, the options open to a specialist supplier to disseminate his business information product will increase. The tariff structures for Prestel and other systems, of which I have tried to give some indication in this article, may well themselves dramatically alter under the influence of ever-advancing technology and the pressure of competitive challenge from other services.

More fundamental to the future of the specialist information supplier on Prestel might be the fact that the pattern of access to his database is likely to have the characteristic of a large number of frames retrieved relatively infrequently, since it is the speed and ease with which his user can occasionally retrieve the information and the worth of the esoteric information itself which is of value to the specialist user, rather than a steady source of regular data. For a subscription-based service, perhaps having zero-pence frame-access charge, this makes for a good and useful business for purveyor and consumer alike—but it is not the pattern of usage which will best endear the IP to the PO, which presumably wishes to achieve high usage from each frame in the system. All these caveats apart, one must, I think, conclude that overall Prestel is definitely an opportunity in which the specialist business information supplier should try in some way to become involved, provided the tangible investment can be kept quite low ahead of receipt of anticipated revenues.

126

Prestel International

Given the current capacity problem of the Prestel UK Public Service, compounded by the fact that the £1m advertising campaign for Prestel UK mounted by the PO in the spring of 1980 was angled almost exclusively at the 'residential' rather than the business sector, it may be the new development Prestel International which actually affords the best opportunity for the small specialist business information supplier to take part.

Prestel International is a new departure for viewdata which has been developed along quite significantly different lines from Prestel UK. To begin with, the management and co-ordination of the project has been contracted out to Logica, one of the UK's oldest computer systems and software management consultancies, who seem to have profited by the Prestel UK experience in the pains they are taking to achieve with the Prestel International Market Trial (PIMT)—a one-year test service which will go out from a single London-based computer to 300 users in 7 countries (the USA, the UK, Sweden, West Germany, the Netherlands, Switzerland, and Australia)—an entity with a much sharper commercial cutting 'edge' than that achieved so far with Prestel UK.

The PIMT has therefore the following important features which contrast markedly with the Prestel UK situation:

- the database contains 'business only' information, and is split into a 'public' area, freely available to all 300 users; a 'private closed user group' area, to be accessible to perhaps 100 of the users or an 'in-company' basis; and a 'syndicated closed user group' area, to be accessed by probably 5 CUGs of 20 users each, all recruited on a basis similar to that operated for example by Infolex in the UK, as described above;
- the IPs are by 'invitation only' and, with a few exceptions, are drawn from the most active and internationally known Prestel UK IPs supplying business information;
- Logica is exercising a very limited amount of overall database 'management' (or, better, 'co-ordination'—one hesitates to call this 'editorial control'), allotting distinct information areas to specific IPs, so that the whole PIMT database stands up as a complementary, rather than competing, set of information products, each of a high standard of content and presentation.

It may well be that Prestel International proves to be a bigger success story in a shorter time internationally than Prestel UK is proving domestically. It is of some significance, for example, that the British Broadcasting Corporation (BBC), not so far an IP to Prestel UK, has been invited to participate in the PIMT. Trading under a new name 'BBC Data', the BBC's entry into Prestel International is in fact part of a wider initiative on the part of its Reference and Registry Services Department (which internally controls and maintains a large number of library and information units for the Corporation's use in broadcasting radio and TV programmes with the widest possible production brief) to seek new revenue-earning opportunities in the internationally growing information-services market. The author should declare an interest in this development also.

The information provided by BBC Data to the PIMT (in the 'public' area of the database) is to consist of a unique and integrated database covering 'World Business and Trade Events and Personalities'. This will include an up-to-date 'Diary of World Political and Economic Events', and a 'World Personalities Index', showing details of important trade, business, and technology ministers in the governments of 25 countries in the Far East, Middle East, and Eastern Europe.

The BBC is perhaps generally the UK's biggest 'information supplier' (although it might also be viewed as having many of the characteristics of the specialist business information supplier, too, particularly in its 'sourcing' expertise and resources). It has an unrivalled world-wide reputation for accurate, reliable, and impartial reporting, and these qualities are intended to be carried over to the offerings of BBC Data. It is also of interest to note that BBC Data's involvement with PIMT has internally created a useful liaison with BBC TV's news and current affairs' Ceefax teletext magazine—the PIMT 'on-screen' logo for BBC Data, for example, being designed in conformity with that for Ceefax, as already seen in the UK broadcast on teletext-receiving TV sets.

Conclusion

All in all, then, Prestel could now well have a future as a major new medium, alongside the press, radio, and TV. Only then will this last, classic, joke stand up: Hannibal, Alexander the Great, and Napoleon were reviewing the Russian military might passing through Red Square one May Day. 'Gosh,' said Hannibal, pointing at the tanks rumbling along, 'if I had had those, nothing would have stopped me.' 'Yes,' said Alexander the Great, pointing at the rocket launchers as they rolled by, 'and if I had had *those*, the whole world would have been mine.' Then they noticed that Napoleon was strangely quiet, and, turning round, they saw he was sitting in front of a PRAVDATEL viewdata TV set, punching at the keypad with a faraway look in his eyes. 'Gentlemen,' he mused dreamily, 'if I had had *this*, no one would even have *heard* of Waterloo . . .'

10.   Prestel and the consumer

St John Sandringham

Consumers need information. They need to know what products are available to them at what price. They need to know their rights and entitlements and they need to know who to turn to when things go wrong. Much of the information they need is purely factual, but as life has become more complicated and products more complex, there is an increasing need for evaluation. The bare facts are meaningless to many consumers, or are too many and cumbersome to be useful. In a competitive society there is no shortage of information for consumers. Every manufacturer or producer of any kind is naturally keen to tell the consumer of the existence and virtues of his product and to persuade him of his absolute need to purchase it. Authority too recognizes its duty to inform the citizens of its policies and their rights and duties. The result, for many a consumer, is a state of information overload—that is, having more information about a subject than he can make sense of in his circumstances—or at worst, finding his information polluted by exaggerated claims or the omission of vital facts in a slanted comparison.

Consumers have banded together all over the world in attempts to make their needs known to producers and authority and to protect each other against exploitation. The Consumers' Union in the United States was founded in the 1930s and helped the infant Consumers' Association to get started in the UK in 1957. The Consumer 'Mafia' is now world-wide with a central organization—the International Organization of Consumers' Unions—in The Hague.

One thing which practically all these consumer organizations have been successful at is acting as information co-operatives. They have collected and classified the information consumers want. They have employed specialists to test and evaluate products and services. And they have 'published' the results—usually through the medium of a periodical magazine. There are about 150 national consumer magazines covering more than 50 countries and many more regional or local magazines, bulletins, or newsletters.

But consumer organizations have always sought other effective means of spreading the results of their research. Book publishing is an obvious way of offering accumulated experience on subjects as varied as eating out and doing your own legal conveyancing. Newspapers, other magazines, and the broadcasters have often been glad to co-operate with consumer bodies and have provided a welcome amplification of the message and an opportunity to reach a wide public. The face-to-face interview is perhaps the only way of communicating some kinds of information and advice. It also provides the opportunity for a

129

degree of personal problem-solving not possible in other media and opens consumer advice up to people in both developed and developing countries who are unlikely to absorb it through the more literary means of communications. Advice bureaux have proved an effective way of giving consumer advice on a person-to-person basis.

Against this background of development and innovation, it is not surprising that consumer organizations in many parts of the world are taking an intense interest in the potential of viewdata/videotex systems. Videotex promises first of all to be an efficient means of information distribution. Although costs must look prohibitive to many people in 1980, the trend is on the side of electronic distribution. The cost of printing the UK magazine *Which?* and distributing it by post has been increasing over the past five years at around 25 per cent per annum on average, and shows no sign of abating. The true cost of telecommunications services, computer hardware, and domestic electronic equipment has, on the contrary, declined when compared to the retail price index. An established Prestel service with receivers in a million or more homes could be a cheap way of providing consumers with the information they need.

Secondly, providing information via videotex can be self-financing. The consumer pays for the information he requires. He thus pays the piper and can call the tune—an important principle for organizations which exist only to serve consumers and are jealous of their independence from producer interests, including government in its role as a provider of services.

Thirdly, there is the editorial opportunity which videotex presents. Providing information which is more up-to-date and more closely tailored to the individual consumer's needs has been a constant challenge for consumer organizations. With videotex, fast-changing information like shop prices and availability can be up-dated at the frequency the consumer is prepared to pay for. The practical limitation is the cost of collecting the information, not that of distributing it. Instead of being forced to buy the package of reports on a variety of subjects dictated by the needs of the majority and the demands of printing schedules, he can selectively buy the information he needs when he wants it. It is also possible to provide him with information like local prices, 'best buys' which take into account his own minority requirements, and a suggested course of action which takes into account his personal circumstances. The best buy for someone who insists that his new fridge is silent, has automatic defrosting, and fits the space left by his old one is not likely to coincide with the best buy calculated on average requirements. Similarly, the value of using a limited 'window' to look at your own small part of a complex set of administrative regulations has been demonstrated on Prestel using entitlement to state maternity benefits as an example.

Of course, the videotex medium has limitations for consumers. The most obvious are the small amount of information which can be conveyed at one time (so it is difficult to compare products for a wide range of parameters), the limited display possibilities (so you can't see a realistic picture of the product under discussion), and the lack of portability (so you can't take it with you to the shops). Some of these may be mitigated or, in future, overcome. They do, however, suggest that videotex is more likely to find its place as a useful addition to the range of media used for conveying consumer information than as a total replacement.

So much for the potential of videotex perceived by consumer organizations. What does the consumer think of it? Early research by Consumers' Association with groups of people of different socio-economic background produced a widespread acceptance of the concept of calling information on to a TV screen. The development of such a system was seen as inevitable but the extent to which it was welcome varied. One difference which was consistent over some 13 group discussions with 10 to 12 people participating in each, was that those from the top of the socio-economic scale—the professional people, senior managers—tended to be wary of the potential snags, while those lower down the scale—junior managers, blue-collar workers—tended to be more excited by the new opportunities such a system would present.

The snag which united all social groups, however, was that of high initial cost. These group discussions took place before the Prestel public service and before any tariffs had been announced. The members of the groups were, however, intuitively correct in assuming that both the cost of the modified TV equipment and the cost of usage would be a major deterrent to people having Prestel in their homes initially. The full advantages of videotex for consumers are only likely to be realized in a true mass market where the costs can be spread over large numbers of buyers and subscribers. It is vital, therefore, for consumer information providers that any videotex system in which they are involved is firmly aimed at a mass market, and does not get side-tracked into a high-cost, low-volume, specialized market backwater. That means in practice that the system as a whole must appeal to the numerous families in the middle market, even if they are not the immediate purchasers of the hardware.

Another snag foreseen by many of the potential users who took part in this research was the difficulty of establishing the status of the information on a viewdata system: 'It depends where the information is coming from . . . I think that's the crux of the matter as far as I'm concerned: who's putting the information on?' Doubts were expressed about how up to date and how reliable the information would be. These doubts were coupled, particularly in the middle-market groups, with the question of the system being used for advertising. Clearly, information providers in general are going to have an uphill task to win the confidence of consumers and that task will be more difficult if either the data is seen as predominantly advertising, or if consumers find they cannot easily distinguish between 'advertising' and 'information'. Often the question of advertising was allied with cost: 'Would you feel happy about paying for a pamphlet of adverts?', and 'They won't reduce your phone bill—you're paying for the privilege of looking at their ads.'

Generally, however, the potential of viewdata/videotex as a buying aid was recognized: 'If you want to buy something it's terrific'; 'I'd probably look here rather than magazines; it's easier'; 'Certainly if I was going out to buy an item I'd look it up—because it's on tap'. This wasn't uncritical acceptance. The information obtained would still be checked against the experience of friends, for example. There was also the feeling that the cost of using videotex would not be justified for cheap items. 'You're going to spend 20p to find out something's gone down 1p.' The extent to which viewdata/videotex allows the 'perfect market' to be approached will depend on how successful information providers are at giving information which is comprehensive, up to date, and local. There was even a school—particularly noticeable at the top end of the market—which thought that comprehensive buying information on viewdata/videotex would 'take the fun out of buying'. Some even thought it would encourage laziness: 'It comes into the same category as remote control television. I mean, who needs it unless you're disabled.'

However, the evidence for the need for consumers to 'shop around' is incontrovertible. You don't always 'get what you pay for'. *Which?* magazine in the UK can point to discrepancies like £75 difference between the highest and lowest prices for the same Sony TV, and a possible saving of 18 per cent in the price of a basket of groceries just by changing shops. E. Scott Maynes in the USA has compared price and quality systematically over a number of articles to demonstrate what large departures there can be from the perfect market there. He postulates the need for an information system which would allow consumers to make an informed choice. Videotex is a potential answer to that need.

The extent to which Prestel can answer the consumers' needs was to have been tested in the market trial of the system starting in July 1978. This did not produce the quantitative information hoped for, partly because of the slow delivery of Prestel receivers and partly because of the British Post Office's failure to implement the software which would collect the records of individual calls to the system. All the indications are, though, that consumer-buying advice is one of the more popular topics on the system. *Which?* has been consistently among the leaders in the league table of information providers on Prestel, and the most popular sections of the *Which?* database have been those giving advice on buying things—whether it be a new car, a washing machine, or a restaurant meal.

There is also now evidence of interest in the use of 'response' frames to allow Prestel users to order goods off the screen and complete the transaction by keying in a credit-card number. A number of retailers have now progressed from listing goods and prices to offering them for sale via Prestel. They join a small band of publishers and specialist mail-order clubs who have been offering goods in this way since the facility was introduced. Buying goods by pressing buttons on Prestel is undoubtedly an added convenience, for the consumer. Some will see in it sinister implications such as a future generation, mekon-like, unable to use its legs to walk to the shops. And there are, of course, serious concerns such as the need for vendors to spell out accurately and fully the terms of any offer they expect consumers to respond to on Prestel. Another concern is the possible exploitation of the natural attraction such a system has for children, by advertising goods for sale specifically to them. These are all areas which consumer organizations will be monitoring as videotex develops.

However, there is little doubt that Prestel is set to become a market place. Goods and services will be advertised and sold, and that would be sufficient reason for consumer organizations to be involved even if the system held no other attractions for them. The Prestel user will need the independent advice and help he has come to expect in other media. All this makes one huge assumption: that Prestel will in fact find its own market among consumers. Given the high cost of receivers initially and the scepticism about their need for such a system among higher-income groups, the way to that market must surely be through communal use. The bulk of consumers are likely to have the opportunity to use Prestel in the early years only if sets are widely available in public places: pubs, shops, railway stations, libraries, advice bureaux and, dare one say it, Post Offices.

11. GKN: an experiment that failed

Mervyn Grubb

I feel somewhat of a hypocrite extolling the virtues of Prestel, conscious of the fact that I have rejected it after a nine-month trial period. But there is a valid explanation for this action which in no way detracts from the developments bound to stem from this new communication medium.

Let me describe the background to our association with Prestel. In May 1978, I read about Prestel in the *Financial Times* and became excited about its possible utilization in the marketing and purchasing fields, and I asked our O & M team to investigate its potential—for us. Immediate reactions were 'Christ—another mad idea of the chairman's'. However, we progressed from that point and took a decision to proceed with a trial in July 1978.

My interest revolved around two aspects—marketing and the communication of stock information. We concentrated upon this second issue and I must explain the problem.

My company is an industrial fastener distribution company with 32 stocking points, a range of 90 000 items, a slow stock turn, and an enquiry pattern that could range outside of the huge portfolio already carried as standard stock. Our inventory was part computerized but with no live communication of stock availability. We believed that Prestel would help us with three problems:

1. To provide swift stock-availability information for a multi-location distribution service company.
2. To be able to provide options as an alternative to the item the customer specifically required.
3. To find a new and stimulating solution to the age-old practice of using manual stock-card systems.

As most marketing considerations and developments relate to products, range of types produced, and availability, it will be of interest to know something about the actual lead time from commitment to implementation of a Prestel programme. The first job was to set up a small committee with a project director, a co-ordinator, and some workers. Seperate requirements which were identified related to installations, demonstrations, and publicity, Post Office support, file development, contracted services, and finally file update. At peak involvement, fifteen people were concerned with the project.

Planning commenced on 1 July 1978 and staff selection, time scale, expense budgets, individual responsibilities allocated, and a master plan were prepared within 8–10 weeks. We were so confident of success that almost simultaneously we commissioned file design, the form of presentation, colour coding, and the whole programme was put together in some sixteen weeks.

134

It would not be unreasonable to allow for a gestation period of six months—that is, the time taken from the decision to proceed to the initiation of live trials. We coined the name 'Telestock' to indicate the manner and purpose of our initial use of Prestel. We allowed a period of 4/5 months to launch the project from the basis of knowledge we had then acquired, and split the programme into three specific parts: (1) system implementation, (2) branch installation, and (3) trial assessment.

1. *System implementation.*
   (a) Liason with the outside agency to set up and test our stock file on Prestel.
   (2–3 weeks)
   (b) Modifications and adjustments to the file after preliminary test. (1 week)
   (c) Devise indexing systems for access by the system structure (menu approach). Introduce an additional 'manual' index for rapid access to stock information, by-passing the 'menu'. (2 weeks)
   (d) Liaison with computer centre for the regular production of file update tapes. (current)
   (e) Discussion with computer centre regarding the viability of the future use of on-line transmission of data to Prestel. (on-going)
   (f) Provision of a single committed telephone line for the system at each chosen branch location, to permit uninterrupted use of the terminal, and an accurate method of monitoring of costs (including measuring devices if appropriate). (3 weeks)
   (g) The modification of TV sets to lock-out normal television programmes, subject to the continuing facility to utilize cassette-generated demonstration material. (2 weeks)
   (h) Investigation of cut-out devices on Prestel handsets, to ensure branches do not remain 'on-line' when information on 'Telestock' is not being accessed.
   (2 weeks)

2. *Branch installation.* In parallel to some degree.
   (a) Production of pre-installation check list. (1 week)
   (b) Pre-installation visits to define the TV set and jack plugs location at each branch.
   (c) Brief introduction at pre-installation stage to branch management.
   (allow 2 hours)
   (d) Provision of wall-mounting TV support brackets.
   (e) Preparation of a branch presentation.
   (f) Preparation of 'script' for training presentation to branch staff.
   (g) Programme prepared for branch manager presentation.
   (h) Development of training package and manual. (2 weeks)

3. *Trial assessment.* To cover:
   (a) Transmission of data via telephone lines.
   (b) Reliability of terminals/modems (a maintenance record to be kept at each location).
   (c) Accuracy and reliability of the updating of telestock information using preview.

135

(d) Appraisal of the handsets, their operation and design.
(e) Cost of telephone calls per location.
(f) Prestel access charges per location.
(g) Terminal rental charges per location.
(h) PREVIEW running costs (PREVIEW is a software program for reformatting already computerized data onto viewdata systems).
(i) Appraisal of the indexing methods, development of any modification to increase access efficiency.
(j) Interview customers for their opinion of Telestock. This to be carried out independently by an outside agency and GKN personnel.

If any potential user is concerned with the presentation of statistics, dimensions, quantities, prices, then programme 'updates' or revisions will be necessary. We utilized the 'bulk update' package specifically designed for this purpose by langton software, and this 'PREVIEW' package was contracted through Mills & Allen Communications for the trial period.

If the use of Prestel is mainly to give stock-availability data then, in the interests of economy and customer benefit it is highly desirable to carry out an analysis of the stock range to determine the popular items, value ranking significance, and frequency of demand. To adopt the Pareto law philosophy—the 80–20 rule, but grading the stock a, b, or c according to demand importance, it is an undisputed fact that a small percentage of the goods offered will make up the bulk of the sales revenue, and a disposition of this nature is not uncommon.

A items—5 per cent of range—60 per cent of turnover
B items—10 per cent of range—15 per cent of turnover
C items—85 per cent of range—25 per cent of turnover

Having categorized the stock thus, it should only be necessary to update the A items weekly, B items fortnightly, and C items monthly.

At the time we went 'live' we were giving stock availability and quantity details on a range of 5000 fastener items involving 250 frames. At the time we launched our trial programme our options of site location were limited geographically, to some extent, for the BPO had only one main-frame computer at the disposal of Prestel and operating distances from it were controlled accordingly. Nevertheless, we spanned 150 miles (240 km) from Nottingham in the East Midlands to Slough in the South and at peak we had 11 terminals in use. Ten of which were rented and one purchased.

A factor vital to the successful development of Prestel for marketing purposes is to ensure that adequate staff resources are available. If in-house strength is not sufficient in numbers or adequate in skill then it is best to employ trained contractors, and there is a plethora of them. We allocated to our project for the course of the trial one senior O & M man with two trained assistants. The project director was the purchasing director and the co-ordinator a marketing executive with high clerical skills—and this is an unusual combination. An information text was produced for the staff and a training period was provided for all those involved. The duration of it was of two days with an O & M assistant allocated to each branch location for a week. Staff quickly familiarized themselves with the Prestel system, and confidence and assurance in its use was rapidly gained. Over the whole period of the trial, some 40 staff were exposed to the use of the Prestel system and the results were monitored, two questionnaires were produced and the following is a summary of the most pertinent questions and answers.

Question	*Response*
1. Do you think it was a good idea to do this 'Telestock' experiment?	21 Yes 6 No
2. Were you properly introduced to the system and given sufficient training?	26 Yes 1 No
3. What has been the most difficult part of using 'Telestock' in everyday orders and enquiries?	0 'Learning how' 24 Getting connected 3 Working the keypad 0 Getting the set from someone else
4. How often would you say Prestel was engaged when you wanted to use it?	0 Every time 9 Every other time 15 Sometimes 3 Rarely
5. How many times do you use the set per day for frozen stock items?	12 1–5 11 6–10 3 11–15 1 more often
6. Has it been useful for anything other than 'Telestock'?	10 Other proper uses 9 Other Prestel 1 TV 7 Nothing
7. Can you get Ceefax (the BBC's teletext service) on this set?	9 Don't know Ceefax
8. Please rate the time taken to find an item once you are connected to the computer.	0 Very slow 6 Slow 9 Adequate 9 Fast 2 Very fast

I would like to summarize quickly the test results, based upon personal observations. The inside sales staff was re-vitalized: I could detect a new sense of animation and purpose because different skills were demanded of them and, above all, they had a feeling of personal involvement in a fresh and as yet untried marketing device. As I judged it they took more pride in their work and enjoyed the benefits that Telestock gave them in talking to customers about their requirements. Unanimously, they were enthused by the experiment as it clearly demonstrated the resolve of top management to introduce ultra-modern systems of communication and information. Our customers were impressed by:

1. the obvious ability of the organization to produce a new and integrated stock information system;
2. the ability to see a composite stock range at a glance and evaluate the scale of investment implied. It created a sense of security in us as a supplier, and was obviously regarded as an insurance factor.

Suppliers to whom, we gave identical working demonstrations, had similar impressions and were even more anxious than previously to collaborate with us in schemes to our mutual benefit.

We prepared a questionnaire for completion by the staff and as the answers are illuminating a pertinent summary follows. In whatever manner you employ Prestel and for whatever purpose, very full detailed briefing, as well as training, of staff is essential. We prepared two papers for our staff—one describing Prestel as a concept and the other describing our objective in devising Prestel 'Telestock'.

Despite our later rejection of Prestel, we are maintaining a presence in the sense that we retain two sets with a view to developing new marketing and selling approaches. So why did we eventually reject Prestel? We concluded that an ideal use of Prestel was precisely the one we had experimented with—stock information. But the prime requisites for maximum cost and profit benefit are:

1. The nature of business should require rapid and accurate information, related to precise delivery information.
2. There should be limited stock variety—1000, items say.
3. Items should be of relatively high individual value, above £10 each, say.
4. Items should be of a popular nature.

Our practical application was the reverse, in that we featured:

1. Unpopular, slow-moving articles.
2. A wide range—12 000 items (selected from a total stock of 90 000). It should have been all or nothing.
3. Speed of delivery was not essential in that the customer was usually so relieved to locate an unusual demand item in stock that he could wait a few days for delivery.
4. Our prices are not measured in cost per unit but cost per 1000 pieces—incredibly low—with an average of £2–£3 per 1000 for machine screws.

Set-up and development costs

In a distribution business, overheads have to be low, and our actual staff cost was just over £9 000, and with 50 per cent overheads, say, £13 500.

Operating costs

Excluding internal staff salaries, our expenses over the trial period amounted to £29 000 and we feel that this cost was fully justified in the light of: (a) the experience we gained, and (b) the sale of £300 000 of goods we might not otherwise have achieved. The approximate cost of one terminal for a year will be in the region of £1000.

For a national set-up of, say, 30 terminals and 2000 frames, with regular updates, weekly and monthly, it works out at roughly £45 000 per annum or £1500 per terminal with full update and revision services.

So far I have only given information about practical experience and costs in running a multi-point viewdata or Prestel service. But what we have offered to our customers is a marketing service—a static marketing service related to stock information. For the reasons I have described earlier, this particular use by us of Prestel is not the most effective; but having abandoned our first trial application we mean to exploit our experience in this field and, hopefully, chart new paths in the marketing-and-selling jungle.

A Prestel terminal costs about £30 per week for as many hours as subscribers care to use the service: but shall we say 50 hours. So we have 60p and hour sales cost for an effective, accurate, constant colourful sales aid, always assuming that your customers or potential customers are, or will be, patrons of the Prestel system. How can we urge clients in our various industries to patronize the system? By spelling out the advantages to them and us:

1. The opportunity of single sourcing based upon confidence, because the customer will have access to the supplier's stock file and know the extent of the range and depth of stocks carried (no 'brass plater' will be able to live in a developed Prestel world). Single sourcing is the phrase used to describe the principle of buying all your needs of a particular material or group of materials or components from one source. 'putting all your eggs in one basket'.

   The advantages of this practical philosophy are:
   (a) Much lower purchasing costs.
   (b) Reduced progressing costs.
   (c) Less paperwork.
   (d) Better service and quicker supplies.
   (e) Better or lower buying prices because the purchasing power is not fragmented.
2. Regular price update facility.
3. If required, back order or forward order intelligence can be provided.

As a straightforward sales medium Prestel is used to:
1. Describe the main activity of the potential supplier.
2. Provide important statistics such as
   (a) size of factory;
   (b) number of people employed;
   (c) type of production or plant;
   (d) trades supplied;
   (e) industry connections.
3. Supply detailed information about inspection procedures and quality control, thus partially removing the spectre of product liability.
4. Give an indication of financial substance by showing rounded figures of fixed assets, working capital, cash flow, etc.

5. Quote selectively, as a reference to reliability, a number of established customers, who presumably are satisfied with the suppliers' service.

Other themes related to an industrial sales or marketing activity would be the ability to advise speedily upon:

A design change.

A price change—so important these days when two price 'hikes' a year is the accepted 'norm'.

New products to be introduced.

Installation or maintenance advice upon the use of, for example, a standby generator; a heat pump; a central-heating boiler; a sauna bath.

A basic intoduction to new buyers of the products or services sold, i.e., a short course in product knowledge to embrace such dissimilar items as: electro–plating; fabric dying; the purchase of ferrous or non-ferrous metal; engineers' small tools; adhesives; abrasives; fasteners.

There are a vast number of selling 'ploys' to be used—purely to encourage business. For example, it is our intention next year to refine our stock file and to provide stock information data calculated to satisfy 85 per cent of normal stock demands. We will say to a potential customer, 'We will provide a terminal at 50 per cent of the total cost for a year, and it will be our silent salesman for you will not be serviced by a salesman; or 'We will provide a terminal at no cost to you but after a year a charge will be introduced geared to the value of the business generated. If sales to you are substantial, there will not be a charge'; or 'You pay for a terminal, for it has other uses related to other sourcing problems, but we will give you a Prestel discount, say 5 per cent off statements paid promptly or a yearly rebate on sales from stock'.

But there must be scope for even greater versatility and innovation when considering marketing in the consumer-based trades which tend to be brand-dominated. This will probably follow when more sophisticated and refined terminals are available with cassette options, as they surely will be in the future.

Let me throw out a few ideas of potential uses:

1. *Men's wear.* Audience: male householder.

   A leading house has local and regional stockists of high-class men's wear. Even a substantial outfitter can only carry a limited range of styles, fabrics, designs, and sizes. How convenient it would be to be able to switch on the 'in-house' Prestel and show the interested client the full range of XYZ brand clothes in terms of style, cloths available, sizes, prices, and delivery time.

2. *Training organizations and professional institutes,* perhaps specializing in general management studies.

   Audience: career masters, personnel officers on business organizations, government training officers, with the unemployed using Prestel in the local library, possibly provided by an enlightened government.

   The introduction would specify the services provided, and give a breakdown of business functions covered. Individual files would have a frame which described the various job opportunities and prospective earning levels and would proffer a variety if courses. The final frame or page would outline the syllabus of the particular course, dates available, where located, and the cost.

Possible clients would include: the Institute of Purchasing and Supply; the Institute of Marketing; the British Institute of Management; Institute of Chartered Secretaries; etc.

A current user is Glasgow University, and the building and construction industries intend to employ Prestel, in these terms: 'By making readily available a mass of construction information between designers, quantity surveyors, contractors, and suppliers.' There is great scope for use of Prestel in hardware distribution, and it could with adaptation be used in food distribution.

3. *The hardware field.*
Audience: captive clients—part of a symbol trading concept. Numbers can range from 250 to 2500.

The purposes here could be the following:
(a) Introduction of new products on offer.
(b) Relevant statistical data on all high-priced items offered by the distributor and not normally stocked by the independent trader because of capital investment.
(c) Monthly offer or promotion.
(d) Price changes.
(e) Model or pattern changes.
(f) Delivery state of 'top of the pops'.
(g) Seasonal reminders.

4. *Wholesale cloth merchants*
Audience: private tailors, cutters, men's outfitters who make suits to order. Estimated at 7000–10 000.

The purpose would be to describe services and material offered, to include terms of business. The file would comprise many sections devoted to types of cloth, weights available, colours, shades, twists, weaves, with prices updated as required.

The whole of the marketing, purchasing, and sales functions are going to be changed by this new contender in the communication field—Prestel. What current activities and occupations are affected by this new force?

Activities which can be affected are: direct mail; TV advertising; local commercial radio; bus advertising; newspaper advertising; trade-paper advertising; exhibitions; conference contractors. Occupations which are likely to be prejudiced include: door-to-door selling; some organizational selling; bill posting; progress chasers (critical path analysis); junior purchasing staff; night-school teachers.

I am a disciple of Prestel and it really spells *profit* in the future but the rewards will not be immediate. We tend to expect too much too quickly from any new experiment and the best of Prestel is yet to be seen. It is a new dimension in the measurement of marketing effectiveness and its application will spread.

The Post Office has produced a paper: 'Mail order and Prestel', and I quote its concluding comments. 'Even with the present technology the following advantages will be available.

1. Direct ordering by agents through the Prestel system.
2. Substantial savings in staff and equipment through reduced VDU requirements in company offices.
3. Details of out of stock items available to agents before orders are placed.
4. Greater flexibility in sales offers and special purchases.'

There are a myriad of job occupations likely to be affected by Prestel in its current form, and as it becomes more flexible and adaptable there is hardly a profession which is likely to remain impregnable. Certainly the development of its use and the variety of applications for which it can be employed will have an enormous impact upon buying and selling. These essential business functions will be changed out of all recognition, and the humble 'traveller' of today will certainly have to seek some alternative form of employment. Setting aside the fundamental aspects of selling, the traveller is normally expected to be a provider of verbal services—the price and availability of the products or services he has to offer, the technical and quality aspects of the product he wishes to sell. His sales after-service includes the provision of information upon outstanding, uncompleted orders, on changes in design, or modification of existing standards. In the future, access to this knowledge will be provided by a Prestel set—quicker, more accurate, always available—at a fraction of today's sales' cost. It has been estimated that a salesman on the road has an all-in annual cost to his company of around £10 000, and based upon experience the average cost of a customer call is £5—wasted if the customer is otherwise engaged.

And what a cheap, reliable, and constant servant Prestel is. No industrial-relation problems, no wage or salary disputes, it is an excellent time-keeper and occupies the minumum of office space. The salesman's future is threatened. Prestel can provide for the buyer all the product information he is likely to require—availability, price, technical specification, appearance. Every forward-looking distribution business of the 21st century will have to offer, as a prime necessity, the Prestel facility.

Part Four

How to do it

```
P R E S T E L
PRESTEL EDITING SYSTEM                        op
  Input Details -

             Update option  o

Pageno  2481                  Frame-Id  a

User CUG              .       User access  y

Frame type  i                Frame price  0  p

             Choice type - f

Choices
  0-   248             1-   5318
  2-   531789          3-   2467
  4-   24857           5-   2480
  6-   24800           7-   248766
  8-   53177           9-   269        ▮


KEY  #  TO CONTINUE
```

Editor's introduction
Electronic editing: Database and page design[1]

The most distinctive and most unfamiliar characteristic of Prestel and other viewdata systems is the structure of the database. Many people, especially those reared in the world of newspapers and magazines, see viewdata as just a sequence of pages. Even those familiar with teletext tend to see viewdata as just a different way of displaying a larger number of screen pages. Yet the design of the database structure is the most fundamental activity involved in setting up a viewdata information service. Without a good database design, the information service, whatever its information content, is dull, lifeless, hard if not impossible to use, and perhaps most importantly, fails to exploit the distinctive features of viewdata to create a distinctive product.

Almost as bad, there is a half-way house of computer and other experts who understand that these viewdata services have a structure, but have interpreted that structure, on the analogy of other structures of computerized information systems, as a 'tree structure'. Articles and lectures abound in references to the tree structure. There is no tree structure: it does not exist, except here and there by choice. It is a pernicious concept that is almost as bad as having no conception at all—maybe even worse, since with no preconception the mind is at least free to grasp the reality.

The reality, and indeed the basic peculiarity of all viewdata systems, is that any page (a page being a single screenful) can have attached to it up to 10 automatic links or routings to any other page. Thus the user can be told, when looking at page 10, that by pressing 1 on his keypad he will be taken to page 531 078 542 (or rather, to the information on page 531 078 542, since the user is not interested in the page number as such), and that by pressing 2 he will be taken to page 24 807 (or the information on it), and so on. Only slightly less important as a peculiarity is that the user can get to any page at any time by simply keying the page number. Thus if he knows that the information he wants is on page 24 807, he keys 24 807, plus two control keys, and there he is.

It is up to you, as the editor or manager of the information service you are offering to the user, to make what you will of these two characteristics. Whenever a page of information is first created on viewdata the editor working on it, or operator, call him or her what you will, first has to fill in a 'shadow page' of

[1] This is an expanded version of part of the study of Prestel written by the same author under the title 'The electronic bookstall' and published in September 1979 by the International Institute of Communications in London, to whom acknowledgement is made.

instructions that the computer attaches to the page of information. This shadow page is the permanent companion to the information page, although invisible to the user. It is, as it were, the dark side of the moon. On this shadow page are entered the page number, page price, whether it is for general public access or restricted access only, and finally the routings or links for the information page, from 0 to 9. The editor may fill in all, some, or none of these, at his choice. But his choice it is, and the effect of his repeated choices, page by page, is to create a structure to the information, that is external to the information itself, but whose objective is (or should be) to aid the user to find what he wants.

Each page on Prestel has a unique page number, from 0 to 999 999 999, and in theory each of these has sub-pages, or frames, denoted by the letters of the alphabet 'a' through to 'z'. Hence the theoretical capacity is 26 times 999 999 999. In practice, the first Prestel computers had a capacity of 250 000 pages, about 150 000 of which were in use by 1980. The way that Prestel is organized in practice is that the Post Office itself (i.e., the system operator) creates a series of master indexes to the total database, beginning on page 0: and the individual IPs, of which there are over 100, are given (or rather leased for money) three-digit numbers as their individual master numbers. Hence Fintel starts on page 248, the Birmingham *Post and Mail* on 202, American Express on 269, and so on. Three-digit numbers are Prestel's status symbols: without one, you are known, a trifle derogatorily, as a sub-IP, since you have to use pages provided by someone else who has got a three-digit entry-point. With a shortage of pages such as occurred in 1979–80, there developed a sort of black or secondary market in spare pages.

All the information provided by a particular IP sits 'underneath' his three-digit number, and in practice most IPs have tended to divide pages between index pages, which tell the user what he will find and the numbers to press to get it, and information pages, which are the user's final goal. There is no technical reason why this distinction has to be made, since, as we have seen, every page can have both onward routings and information. But it is a convenient distinction. One problem, however, that the creators of large databases have quickly found (those, say, of 5000 pages or more) is that the larger the database, the greater the number of index pages as a proportion of the total, since the user may begin his enquiry in a number of different ways, and may wish to pursue it in a number of different ways. For example, cinemas may have to be indexed alphabetically; by location; by price range; and by hours of showing; and the films showing may have to be indexed alphabetically, by type (horror, cowboy, art, etc.), and by date of release (new or old) or even by country of origin.

Unless you cater for the many possible motivations of the enquirer, you will quickly lose him, or not get him in the first place, since he cannot easily find what he wants. It is not uncommon to end up with as many index pages as information pages; or indeed more. Now it is becoming increasingly rare to charge for index pages—how can you charge the user for discovering that what he wants is not there? As a result, half your pages may be carrying all your costs as an IP.

As one answer to this problem, some IPs have consciously tried to abolish the distinction between index pages and information pages, as they are technically

146

free to do, and to give information on all or most pages, simply varying the amount according to the number of routings that have to be given as well. That way, all or most pages can be charged for, and the user is spared the often tedious business of wading through large numbers of index pages before he gets to something interesting.

The disadvantage is that the viewdata page is so small. It has only 960 characters, of which only 858 are in practice usable, given that the top and bottom lines are reserved for system use and that the colour coding at the beginning of each line, except for white, takes up a character space. So even with only 1 or 2 onward routes displayed and described (e.g., Press 1 for horror movies) let alone 4 or 5, the amount of space left for actual information gets increasingly constricted. If you are trying to give value for money, especially on high-priced pages aimed at the business market, you cannot give over half the page to non-information content. Each IP must find his own balance between these factors—as long as it is realized that there is a choice, or range of choices.

At this point, it is necessary to step back and ask, what is the overall structure of all these index pages, not to mention information pages, that you as editor or manager of a substantial database are creating? After all, neither the user nor indeed the editor ever sees all the pages at once, or even at one session. It is not like flicking through the pages of a 40-page or even 100-page newspaper, where you can see the overall structure of the product almost at a glance. Unless there is a consciously designed architecture, preferably expressed on paper, pages get lost, the editor does not know where to put anything, let alone where to direct the user, and the user does not know where to find anything. This is why, incidentally, it is not usually possible to generate either random or even sequential page numbers on a computer-to-computer transfer of information to viewdata. The need for an explicit structure makes it important that many sub-numbers remain the same (e.g., the page number of the main local cinema, which customers get used to) even though the quantity of information may vary (e.g., how many movies, of what types, are showing there this month).

As we have said, computer people, famliar with the notion of a tree structure, have also seen videotex as a tree structure, or rather an inverted tree structure. They see it, in its crudest form, as a system in which the user is first offered an index with 10 general choices and, having made his first decision (e.g., business information), is then offered a second index, a little less general, and makes a second choice (e.g., market prices). He is then offered a third index, and makes a third choice (e.g., share prices). A consequence of the way that Prestel is organized, and therefore a peculiarity of it, is that the user is then offered yet another index, showing the various organizations providing share-price information, and he makes a choice (e.g., the Stock Exchange). He is then confronted with the indexes of the Stock Exchange material itself, created by the Stock Exchange: these could be, first, a list of industry sectors, and second, an alphabetical list of company names, and only after making these further choices does the user eventually get his share price.

This quite realistic example shows that the user has to go through six indexes to get to the end information, and illustrates the 'index page overhead'. It is also the reason why there is a strong move to give the user the exact page number of the information itself, either by advertising or in the form of directories, rather like telephone directories, that he can look up, so that he can key the page number direct. It also raises the interesting question of whether the user will really be conscious of the transition from Post Office indices to IP indices, and whether he will really perceive the system as a single, Post Office information system, whatever the formal arrangements. But once a user has penetrated into a particular IP's database, one object of that IP's indexes and routing will be to keep that user inside his database, rather than have him wander off elsewhere.

But the question remains, is this picture of an inverted tree structure a valid one? Is it sufficiently valid to be made, as it is so often, the fundamental information structure of Prestel and of other viewdata systems? The supreme disadvantage of the tree concept in its pure form is that once the user has got to his destination page, what can he then do? Presumably either ring off or start the whole laborious process all over again. He may be reluctant to do either, and anyway it could be disastrous to the economics of the system if the user only looked at one page each time he dialled in. This is known in the trade as the 'mineshaft syndrome'.

So what do you do? What you do is to put on each and every information page details of further related choices that the user can make, and enable him to make them simply by pressing one key on his keypad. Another cinema? Key 1. Another type of movie? Key 2. Another area? Key 3.

Once you start on this line of development, you are in fact allowing the information itself to dictate, within broad limits, the external structure within which it is contained; and that is as it should be. But it also means that different types of information with different users may have different structures. We have identified various different structures, variously labelled cartwheel, ring of rings, lobster pot, chinese lantern, and other multi-dimensional models, as well as the simple tree. But the essential point is that they were originally created by editors working out for themselves how to present their information systematically, interestingly, and dynamically on Prestel. Hence the inclusion of database design under the heading of editing; it is not, as so often presented, the computer manager's province, any more than the tree is anything more than one of a host of possible models or 'architectures'.

Prestel was conceived as a system simple to use, for the ordinary customer. Talk of complicated data structures often draws the accusation that one has destroyed the basic simplicity of the system. But the truth rather is that to attain simplicity for the customer, a truly desirable aim, the editor has to evolve complex architectures of information whose complexity is virtually invisible to the customer but whose objective is to cater for the variety and unpredictability of the information needs of even the simplest human being. As with a physical building, the result of complex architectural design is usually a beautiful simplicity of appearance, and therein, we are all agreed, lies the supreme beauty of Prestel as well.

But the question still remains, how consciously do you express these architectures? Having discarded the notion of a tree, do you nevertheless evolve a limited number of alternative models, each laid down on paper, from which the editor/manager of your database can choose, but outside which he cannot stray? Some IPs do this, and it usually takes the form of constructing algorithms or formulae by which one page may be tied to groups of other pages according to their number sequences. For example, it may be laid down in one of the algorithms that given a page number 2487, then choice 0 on the routing always goes back to page 248 (since it is vitally important to give the user a return route as well as an onward route), while choices 1 to 6 always go to pages 24 871, 24 872, through to 24 876, and choices 7 and 8 are left free for future expansion, while choice 9 is an 'optional choice' that may go anywhere, for example, to a related advertisement. Various such algorithms are possible, and exponents of this approach argue that the important thing is not the exact algorithms you use, since they will vary from application to application, but that you should as an IP be aware of the need for them, have the algorithms that suit you, have enough of them to give a choice but not so many as to be confusing, and write them down so that everyone knows what to do and what is going on.

This is the formal but flexible approach, and it suits some. But for others it is still too formal. There must be general rules, capable of applying across large numbers of pages, otherwise the database is and looks a mess. But there may also be a need for a much larger element of discretion on the part of the editors, to allow them to use their imagination and knowledge of the database to create new and unexpected paths through it. Thus there may be (as in the company with which the author is associated) a rule that all pages must normally have at least two and often three routings attached to them: a rule that there must normally be a backward routing (perhaps on choice 0) as well as a forward routing: a rule that there must normally be an easy path back to the front three-digit page: a rule that similar or identical types of pages must normally have similar or identical types of routing, so that distinct routing patterns emerge: and a rule that routing instructions are normally shown in a distinctive colour, at the bottom, at the left or at the right of the screen, as the preference may be. But these are much looser rules than those implied in the algorithmic approach, and depend in the end on the common sense and experience of the editors entering the pages. Clearly, with computer-to-computer transfer of data, strict rules about routing and indexing have to be drawn up before any transfer can take place, but even this does not, as we have mentioned, imply that it has to be a tree structure, so long as it is a regular pattern that the computer software can create and recreate. Whatever your approach, it must be one that strictly governs use of important four-digit numbers (a scarce resource) and controls accurately the number of pages in use, which can otherwise get out of hand—an expensive mistake when every page costs you to rent.

Database structure raises one last and deeply absorbing topic for those interested in presentation of information. It is the role of the front-end indexes created by the system operator, normally the PTT. On most viewdata systems, it is the systems operator who creates the initial top indexes—do you, the user, want business information, travel information, news, entertainment, and so on? After several layers of refining the user's choice, these system-operator indexes then give way to the indexes created by the actual supplier of the information, the information provider. It is worth adding that by this means the user may not necessarily come to the IP's three-digit front page, but be taken direct to some subsidiary part of that IP's database. This is yet another reason why the user may not necessarily be very conscious of the separate identity of each IP, but rather see the whole Prestel service as some sort of unity from which he picks out bits.

But the fundamental question is, how far do these front-end indexes extend? This splits into two further questions. How deeply into the information can these indexes extend without becoming monstrous in size? And how far is it right that the system operator, as common carrier, should influence and control the access routes to the data taken by the user? For clearly, he who controls the indexes controls the movements of the user and controls the commercial fate of the suppliers of the information. Indexing is *not* a neutral academic task, at least not on viewdata. It is not like a book, where the index is added last. The index is the key to viewdata, at least until such time as this index approach is modified by widespread use of directories and later by use of so-called 'key-word searching' that enables the user to ask direct for 'Majorca' or 'BP' or 'Noughts and Crosses' using a full alphanumeric keypad.

At one extreme, the system operator, the PTT, could just index the IPs, with a brief description of what each provides. Then it is up to the user to find out whether a particular IP has anything he or she wants, much like buying or not buying a magazine on a news-stand (hence the frequently used term 'electronic bookstall' to describe Prestel and similar viewdata systems). At the other extreme, under the banner of making it as easy as possible for the user to find specific bits of information, the front index can reach further and further down towards, say, the time of the London–New York *Concorde* flights or the test reports on the latest Mercedes. The trouble is that on this philosophy there is, at least in theory, no end to the indexes, which threaten to swallow up the entire system. For example, on Prestel, is it possible to have a consolidated index of all companies and corporations mentioned? A reasonable idea, which may be implemented, but it means an index covering perhaps 2000 companies (maybe more in future) and half a dozen IPs. The index must take several thousand pages to construct, and the same might be true of a comprehensive index of all motor vehicles mentioned. These figures, spread over a wide and potentially unlimited range of topics, represent a significant overhead on a system that has, at present, an upper limit of 250 000 pages. It may even be that the effort to make it easier for the user in the end simply baffles him, by enmeshing him in an endless pyramid of indexes. (Prestel participants have a fantasy in which the entire content of the Prestel computer is made up of the complex billing mechanism needed to make it work plus an all-embracing index structure, so that the residual information content finally dwindles away to nothing.)

No doubt practice and common sense will dictate a median index policy, and

150

for those who run a dedicated viewdata system (for, say, stock-market prices only) or run a controlled and managed database, not on the common-carrier principle, these indexing questions are less formidable. But just because viewdata is a system that uses indexes, how to index will be a question of some complexity for any viewdata operator. Indexes are, after all, both part of the inner structure of viewdata and part of its outward appearance, its style.

Every medium has its own style. Its style is part of the way in which its audience identifies with it. Prestel is rapidly developing its own style, or range of styles. There are five elements. Two are traditional in any form of publishing: the layout of the pages, and the editing of the text and figures on the screen. But three are specific to viewdata. These are the easy availability of colour; the indexes of routing structures; and the pricing of the information.

The layout of a viewdata page is governed by the format of 22 usable lines, each with 40 characters. It is a very small page, about half the capacity of a normal VDU screen; and it will hold a maximum of about 130 words. But it has to be visually pleasing, or at least not so hard on the eye that you cannot read it. Each page must be self-contained, look well-organized, and have an obvious visual continuity with other pages of the same or related content. Each IP develops his own house style to cope with his own particular information, and the user will quickly realize that (for example) the Fintel magazine (or database, call it what you will) on Prestel *looks* quite different from, say, the Consumers' Association magazine.

In the same way, the editing demands that something useful be said on each page without fuss or flannel in a very short space. Viewdata does not like long sentences or qualifications. Nor must you cram too many words on to the screen; otherwise it looks a mess. Each page has to have a single message, and a clear one at that. Two short paragraphs to a page, broken perhaps by a blank line, or a few columns of figures to a page, are the most you will find, although a lot depends on whether the page is meant for the general or the specialist user.

Nevertheless, it comes down to what a newspaperman calls 'tight subbing'. The use of colour is totally different. It is a problem intrinsic to viewdata. Seven colours are immediately available at the press of a button. The question is, does colour help to convey information better than black and white? There is a theory, which I do not believe, that business users of viewdata will prefer black-and-white sets, and domestic users will prefer colour. Experience suggests that most businessmen also want colour, though preferably on a small screen.

So how do you choose your colours? Every IP makes his own choice. He could, if he wished, have every word and every figure on every page in a different colour, within the seven available (white, red, light blue, dark blue, magenta, yellow, green). He may also replace the standard black background of the TV screen by a coloured background chosen, line by line, from the seven colours available.

The ability to display in seven colours has caused some IPs to change colour almost from paragraph to paragraph, for no apparent reason: and/or to choose jarring or almost illegible combinations of colour to carry their message. In practice, the range of colour available is deceptive, since the deep colours—dark blue, magenta, red—are hard to read, and should rarely be used, apart from graphic applications and for background colour. White is also to some eyes rather glaring to read in quantity. There remain light blue, green, and yellow as colours that are easy to read and easy on the eye.

There is in fact an established 'luminosity scale' that in part explains these variations in legibility, although the fact that, for example, red letters can be put on a white background on the screen, makes it important not to take the figures too literally. Nevertheless, they are instructive.

Colour	Comparative brightness (%)
White	100
Yellow	89
Light blue (cyan)	70
Green	59
Magenta	41
Red	30
Dark blue	11

In short, colour on viewdata must be seen in low key—comparatively few graphics, and comparatively few, restful colours for the text. Nevertheless colour is valuable, if used with discretion. Individual pages can be made more attractive to look at; their information content can be made clearer; a consistent house style can be given to an individual database; the division between the information content of a page and other (e.g., routing) material can be made obvious; and last but not least, colour can be used to make up for some of the limitations of the size of the screen. The use of colour can to some extent take over the use of white space in the layout of the printed page. In other words, colour can be used to differentiate one column from another, one subject from another, one category of information on the screen from another, thus tightening and improving the presentation. Colour changes should never be arbitrary, but themselves convey a meaning.

Apart from indexing, already discussed, pricing is the other peculiarity of Prestel. Each small page has its own fee for looking at it, in addition to the telephone and computer access charges. With a newspaper, you pay one single sum for the whole lot; the same with a book. The BBC has a single licence fee per annum. But with Prestel, each separate nugget or page can have a price fixed individually by the editor, ranging theoretically from 0 to 50p, in practice from 0.5 to 25p depending mainly whether the page is meant for domestic or business use. So the pages have to be worth it. Not only should the indexes not mislead, misroute, or mislay; but the pages of actual information should be up to date, to the point, value for money. Pricing becomes part of the editing process.

Turning the pages of Prestel, you will therefore find five different sorts of pages. You will find pages to read, that is, text; pages of statistics, like share prices from the Stock Exchange or racing results from Extel; graphics, that is, pretty designs to make the thing look more attractive; index pages, those with the numbers 0 to 9; and directory pages, those with lists of addresses, names, holidays, hotels, spare parts, consumer durables, cars to buy, mainly derived from lists that already exist elsewhere.

The bulk of the pages now on Prestel are of this directory type. Some argue that the small page size means that this is inevitable, and that Prestel will be like a gigantic reference book for people looking for very specific bits of succinct information across a huge range of topics. Others argue that unless viewdata can be developed as an original publishing medium for comment, debate, and argument, on social, political, financial, or leisure topics, then it will never make its mark.

Several IPs have made a deliberate attempt to explore the possibilities of authors writing original material for Prestel. The text has to be written, or at least edited, into the special and highly constricted format of viewdata. One cannot let text spill over from one page to the next in random fashion: there is no way that anyone is going to read *War and Peace* on Prestel. The pages are too small, and the screen has too much glare to read a lot of pages in a concentrated manner.

On the other hand, painting produced the miniature as well as the panoramic canvas; literature produced the eleven-syllable Japanese poem, the Haiku, as well as *War and Peace*. Not that there is any poetry yet on Prestel, though I see no reason why not (there is a do-it-yourself rhyming verse game in pretty poor taste). But there are points of view on almost anything that does not need a million pages to propagate.

As to graphics pages, the fact that Prestel offers colour graphics at all is of course itself a distinct feature of the system. But it must be recognized that Prestel graphics are, at the present stage of technical development, very crude in design terms, being made up of comparatively few building blocks of colour in relation to the total screen. Unless and until they are refined to the point of being able to show, say, accurate road routes for motorists, the graphics will be mainly decoration—and people will not for long go on paying real money for decoration. Graphics will have a limited use for some years to come—limited perhaps to title pages of individual databases, advertising applications, and games. Other viewdata systems of different design, and Prestel in the future, may have quite different graphic capabilities.

With different sources, different layouts, different colours, different purposes, it is not surprising that the Prestel pages speak to the reader in different ways. The directory pages, those with lists, have the simple objective of providing facts, dates, numbers, prices. Other pages talk to you in a very conversational way, almost as if having a chat with you; this style is rather like a page in a tabloid newspaper.

There are also advertising pages. We still do not know how much advertising there will be on Prestel, and quite what it will look like. Then there are computers talking to computers. Some IPs have their information on computer anyway, and do a periodic machine-to-machine transfer. Whether the resulting Prestel pages are good or bad is perhaps a matter of taste. Some of the worst-organized material to have appeared on Prestel, as well as some of the best, has been fed in

153

computer-to-computer. On the other hand, in the long run it may be the cheapest and most efficient way to supply viewdata.

For all these reasons, the relationship between the publisher on viewdata and the reader/user is a novel and delicate one. When you tune in to Prestel and start pressing the numbers on your keypad, the IP may be speaking to you in several different ways. He may be saying, in effect, 'I offer you *either* A *or* B *or* C to look at'. That is the simplest version of the choices offered by the indexes. Or he may be saying 'I offer you A *and* B *and* C', where he is inviting you to build up a cumulative picture of, say, a holiday in America or a British company.

Or he may be saying 'I suggest you try A, *then* go to B, *then* go to C', where the IP is deliberately taking the initiative and guiding you by the hand round his 'information stall'. But equally, he may offer you a random collection of items for you to make your own sense of, in the form 'A + L + P + B + Z'. Or he may have a subtle approach, saying 'Read A; if you need more detail, read B; if you need even more detail read C', so that different degrees of interest are satisfied in different ways.

For all these reasons, editing for viewdata is not like newspaper editing, and not like computer programing, but a novel and challenging mixture of disciplines drawn from both.

12. Using an 'umbrella'

Tim Chapman

Definition and industry breakdown

The service industry, or at least the service profession, has been described as the oldest in the world. Being essentially a 'people skill' industry, it is able to react fast to meet the requirements of new markets and it is hardly surprising that it was one of the first to exploit the commercial opportunities offered by the emergence of viewdata technology. For the purpose of this chapter, I define the viewdata service industry as those individuals and organizations which offer, for commercial gain, skills and facilities to information owners wishing to use viewdata technology to communicate that information.

Thus, this definition embraces consultancy, education, software provision, and publishing as well as the supply of 'umbrella' services which enable information owners to subcontract the publishing of their information through viewdata to a third party. It is to this last category, however, that I will devote most attention because it is, in many respects, a totally new activity.

This review of the viewdata service industry, in the interest of charity primarily, examines the following discrete activities: viewdata services on public access viewdata systems such as Prestel; viewdata services offered on private viewdata systems, though this activity hardly exists yet; other services of a more general nature. Once again, for clarity, it is convenient to distinguish between those activities which are concerned with the provision of services *per se*, compared to those where the provision of services is complementary to, or a means to achieve, the publication of a comprehensive information service. It must be stressed, however, that there exist no clear boundaries and, as the technology develops, the boundaries between activities will become further blurred.

Pure service provision on public-access viewdata systems

This activity, which most closely resembles the involvement of my own organization on Prestel in the UK, is to a significant extent, a totally new activity which might be likened to a cross between publisher, advertising agent, printer, and computer bureau. The generic name given to such a service supplier on Prestel is 'umbrella information provider'. It offers the organization wishing to disseminate information on viewdata a relatively inexpensive, low-risk, fast and easy way to achieve their objectives. Alternatively, it offers a range of skills and facilities to help organizations learn about, evaluate, and if necessary assist in the implementation of viewdata information systems. The key elements of the activity are that the service provider is paid for the provision of those services, and his ability to retain his clients' business into the future is largely dependent on his own efficiency and level of service.

Given the present environment surrounding Prestel, the intense interest in the technology by the information industry, and the scarcity of skills and space on the database, it was inevitable that this service activity should have merged. The concept was originally pioneered by two organizations with completely different backgrounds, Mills and Allan and my own, Baric Computing Services, during 1978. Two other names are particularly worthy of note as pioneers in the pure service-provision field: Butler Cox and Partners, for their early consultancy, educational, and market research activities in the UK and elsewhere; and Langton Information Services, for their enterprise and expertise in the development of computer software for the automatic creation and updating of information systems. Many other service providers have emerged, many pioneers in their own fields and some of which will be mentioned at a later stage.

Under the general heading of Umbrella Services a number of distinct activities can be identified:
- The design and documentation of viewdata systems
- The supply of space on the Prestel database
- The creation and maintenance of information systems
- The provision of miscellaneous support services including education, consultancy and research, documentation, processing of customers' information before it is input to Prestel, processing of clients' response frames, and assisting in the provision of terminals.

The typical umbrella client will make use of the first three services, the minimum needed to get information on to Prestel, but others, particularly established information providers (i.e., those organizations which rent their own space on the Prestel database) and organizations contemplating the provision of an information service in the future, will pick those services which best meet their present requirements.

Using a hypothetical example to illustrate some of the services, consider a publishing company which has a superficial awareness of viewdata and decides it needs to understand its business implications. A decision on what action to take will then be made. The company has approached the PO for information about Prestel and the PO has suggested that they should talk to established IPs, including umbrella information providers. As a result of these conversations a service information provider is asked to prepare a brief report on the likely implications of viewdata on the client's industry, identify a number of possible ways the company might use it, and prepare an outline design for one possible system which can be costed and matched against various revenue projections. As an integral part of this consultancy study, the outline system is mounted on Prestel so that the company can get a better impression of its feasibility. The cost for this exercise is around £1500.

The pilot system is demonstrated at the next board meeting and the report used as the basis for a decision to spend up to £10 000 with an umbrella information provider to implement the system. An umbrella information provider is preferred

156

primarily because it is felt that speed is of the essence and the company doesn't want to go through a long learning period. In view of these considerations, it is felt that the costs of using an external service provider are, if anything, marginally lower than using in-house resources, and the company always has the option of doing the work in-house in future once it has more experience of viewdata.

Without revealing the nature of the system, the company approaches three umbrella information providers and gets quotations that confirm that the costing in the consultancy report are in line with the rest of the industry and, partly to preserve confidentiality about a new venture, the original proposal is accepted. The umbrella information provider is asked to present a contract which, after suitable massaging, is signed. During the process of discussing the contract a much clearer insight into the operation of the service supplier and his business is obtained and some important legal issues on copyright are exposed.

The design phase, which lasts six weeks, revolves around the development of a pilot system on Prestel which, although using only some 20 frames, covers the key areas of colour choice, frame layout, routing, and back routing. At the end of this process the company and the service supplier know exactly what the final product will look like.

A detailed dated action plan is drawn up and responsibilities assigned to individuals. Implementation is scheduled to take 4 months. The total system has been broken down into a number of discrete subsystems and the first action is on the company to prepare the information that is to be mounted, subsystem by subsystem. This information is handed over to the service supplier, who is then able to complete the detailed design work necessary prior to input. When the first subsystem has been completely input and checked by the service supplier the company has laboriously to check (proof read) the system on their own TV set. This picks up a number of errors and at the same time minor routing changes are made. This process is repeated for all subsystems and then the whole system checked again. This reveals errors of cross-routing between subsystems and again minor system changes are made. The actual implementation time was $5\frac{1}{2}$ months instead of the 4 months estimated, primarily because of the time needed to proof read the system and check all routings. The cost of implementation was, however, broadly in line with the original estimate.

The company then examines the Post Office index on Prestel in detail and decides what cross-references to his own information are required, leaving the service supplier to deal with the Post Office. Arrangements are also made with publishers of Prestel directories for the insertion of advertisements and other references, and a press launch scheduled. The service supplier is asked to arrange for the supply of terminals for the press launch and to be present on the occasion to deal with any points of detail about the Prestel system.

The general conclusion at the end of the implementation was that the final product was excellent, but that the effort needed to prepare the information and proof read the system was considerably more than had been expected. Although this had led to some friction with the service supplier at times, it was generally felt that the problems encountered were relatively trivial compared with those that

157

would have had to be overcome if the company had attempted to do all the work itself. Also the eventual product was, almost certainly, better designed.

The key decision this hypothetical company took, once it had decided to come on to Prestel, was to use an umbrella information provider. It is important, particularly to the service provider industry, because it is the whole basis for the existence of the industry. An examination of the pros and cons of using umbrella service providers is therefore warranted. The key benefits, insofar as it is possible to generalize, would appear to be:

1. *Speed of implementation.* Prestel is a competitive publishing medium. Because of its openness to all users and its enormous potential, there is a distinct tendency for the first information provider in a given field to discourage his competitors from offering the same product. The umbrella information provider can considerably speed the implementation process, because he has all the basic design and management skills at his finger tips.

2. *Knowledge of viewdata.* Although viewdata system design is very simple compared with the design of conventional computer systems, it is a very new art which is still evolving fast. Also, it is a very visual system requiring a combination of graphic design and information system skills. Without claiming that two years' design experience is sufficient to master the art, a cursory examination of the Prestel database will rapidly show which systems have been built by novice designers. The untrained user may not be able to explain why the system is not as good as another similar one, but he will be able to identify the good ones.

    There are many other aspects of Prestel where an in-depth knowledge is invaluable: who to speak to in the PO; terminal manufacturers and PO plans; knowledge about the activities of other information providers; etc. To keep up to date with this knowledge requires considerable time and effort. Again, a full-time umbrella information provider should have this knowledge.

3. *Resource availability.* Apart from having the required equipment and skilled staff available, the umbrella information provider is likely to be able to make better use of his resources and have the standby capacity needed to meet peak demands and other crises. This is reflected in faster implementation and greater efficiency.

4. *Security.* Dealing with an umbrella service supplier should give the information owner a degree of security he might not otherwise get: the security of a contract, the security of a quoted price, the security that, if he decides to pull out, that is the end of his worries, the security of dealing with a reputable and financially sound organization.

5. *Convenience.* It is much easier to ask someone else to do the work than to organize, manage, and do it yourself, particularly in a totally new field.

6. *Economics.* While it is not universally true that it is cheaper to use an umbrella information provider, there are many cases when this is the case, particularly where the size of the database is small, or the workload is peaky. The umbrella service supplier can achieve real economies of scale in terms of database rental and use of resources, and while he does have some overheads

(e.g., marketing) and the need to make a satisfactory profit which the information owner might not have, he must be able to absorb these in his charges and give good value for money to remain in business. As a general yardstick, if the information owner cannot justify employing at least one fully trained person full time on his Prestel system, he should almost certainly use the services of an umbrella information provider.

Arrayed alongside these benefits are a number of disadvantages which will vary in significance depending on the circumstances of the information owner:

1. *Loss of control.* To a greater or lesser extent the use of a third party lessens the information owner's control. His service supplier may go bust, or wish to pull out of Prestel, or suffer a strike or other natural hazard; he will have priorities which do not exactly coincide with the information owner, or irreconcileable conflicts of interest may arise.

2. *Confidentiality.* The information owner may not wish a third party to have information about his plans or even to have access to the information he wishes to mount on Prestel.

3. *Communication.* There will almost inevitably be some communication problem with the umbrella service supplier; information may get delayed or even lost in transit, meetings may take some time to set up, etc.

4. *Loss of experience.* If the information owner's prime motive for involvement in viewdata is to learn how to design and mount his own systems, then the use of a service supplier will naturally mean some of the learning process is lost, though a reputable supplier is unlikely to begrudge passing on his experience.

5. *Cost.* Given a very large database and sufficient resources to provide for the peak load and standby facilities needed by the information owner, it may well be cheaper to do the work in-house. As a rule of thumb, this point is likely to be reached when the information owner can justify a full-time staff of between 5 and 8 to perform the functions carried out by the service supplier.

In summing up the pros and cons of using the service of umbrella information providers to mount information on a public-access viewdata system, my broad conclusions would be:

- If the information owner requires less than 500 frames it will almost always be very much cheaper to use a service supplier.
- If he needs over 5000 frames it will almost certainly be more expensive.
- If he requires between 500 and 5000 frames his decision will be based on other factors, of which the most important are speed, convenience, and design skill weighing in favour of the service supplier, and confidentiality and loss of control weighing against him.

Finally, I wish to examine the future of the pure viewdata service-supplier industry. Undoubtedly it has a big future and should be able to grow at least as fast as the general growth of viewdata. Undoubtedly the competitive situation will become more severe, particularly from related industries such as advertising and printing/publishing which may feel threatened. At the same time the industry will become more polarized with a relatively few very large service suppliers and a large number of smaller specialist organizations. In general, the full-time service suppliers will grow at the expense of those information providers who are offering services as a convenient adjunct to their other viewdata interests. There will be increasing mobility of information owners between different service suppliers and between a service supplier and a database managed in-house, even though the service supplier may be required to continue to provide certain services. Finally, there will be a continuing move away from manually maintained databases to systems maintained automatically by computers, probably as a by-product of other systems.

Service provision on public-access viewdata systems as a necessary adjunct to publishing

This second major group of umbrella information providers differs from the first in that they provide a service to information owners in order to enable them to publish a more comprehensive information service than would otherwise be possible or financially justified. An analogy to this in conventional publishing is a Yellow Pages directory. The publisher of Yellow Pages has certain information already (name, business types, address, telephone number). By allowing businesses to advertise, the publisher benefits financially, the advertiser benefits (or he wouldn't advertise), and the subscriber has a more useful directory since it contains more information at the same, or lower, cost. Again, this form of publishing/umbrella service activity was an obvious development on Prestel and it is difficult to give credit to an originator, though the Financial Times/Extel subsidiary Fintel (company information), the New Opportunities Press (graduate career advice and opportunities), and Eastern Counties Newspapers (local news, events) were among the first to pioneer such information services on Prestel. Others include PER (Professional Recruitment), Maclarens Publishers (product/service directories), Link House Group (classified advertising), IPC (farming information), Birmingham Post and Mail (local newspapers), Estates Gazette (property advertising), BTV (international investment and trade opportunities).

Allowing for generalization, there are two extreme forms of this activity. At the one extreme the publisher exerts total editorial control over the form in which the information owner's information is presented, and at the other extreme he provides a basic information classification and indexing system, but the content, layout and presentation of each information owner's data is under his own control. The common feature of both is that all information is of a similar type and the information owner pays the publisher for the inclusion of his information. The user accessing the information may also be required to pay, though in a pure advertising application this is unlikely. All systems of this class on Prestel fall somewhere in between these two extremes, with the classified advertising systems (property, jobs, etc.) tending towards the former and the more complex towards the latter. The most useful systems follow the pattern of trade directories, with the publisher supplying wide coverage at a relatively superficial level but with individual information owners paying for the inclusion of additional information with only moderate control applied by the publisher.

This balance ensures that the terminal user is assured of finding some useful information when he enters the system and the information owner is confident that his information is being displayed as part of a system which has an established readership. It is also commercially attractive for the publisher because he has the opportunity to charge both the user and the information owner, his umbrella-service revenues help his front-end cash flow while at the same time increasing the attractiveness of his information, and, with an established readership, he has some means of 'locking-in' his information owners. The only significant disadvantage to this otherwise mutually attractive business arrangement arises from the possibility of conflicting interests between publishers and the information owners who contribute the information that makes their business possible, which unless satisfactorily resolved could result in the establishment of a competing service and the potential collapse of the publisher's business. Despite this relatively remote risk, looking into the future one may confidently anticipate very significant growth in this section of the viewdata service industry, particularly from newspaper, advertising journal and directory publishers, as well as trade associations, independent entrepreneurs, and local co-operating groups.

Other future developments will include the emergence of greater competition as the value of these services to information owners and users gains recognition, and to a lesser extent between pure service and publisher/service providers as some information owners seek to loosen the gentle hold the publisher has on them by seeking other service providers to maintain their databases. At the same time, it is likely that there will be a flow of small independent information providers (maintaining their own databases) under the wing of an established publication. Finally, since the maintenance of such publications lends itself well to computerization, one can expect a fairly rapid movement towards computer-based updating.

Private viewdata systems and the service industry

While the distinction has already been made earlier in this book between a public-access viewdata system and a private viewdata system, it must be borne in mind that a private viewdata system need not be owned and used only by one information owner, nor need access to the information be restricted to one or more closed user groups. One is therefore shortly to see the emergence of private viewdata systems offering services to compete with the public-access systems run by the Post Office or telephone companies as well as private viewdata systems operated by independent service suppliers on behalf of a number of information owners primarily for their own private access.

That private viewdata systems will be operated satisfactorily by the service industry is due primarily to two factors: that the skills required for the

161

management of computer facilities and information are fundamentally different, and that with the falling cost of computer hardware the financial saving to be made by owning the computer hardware as opposed to using somebody else's will progressively fall to insignificance. For this argument to hold water, however, it goes without saying that the service supplier must be very responsive to the varying needs of his information owning clients.

The benefits to the information owner of using a private viewdata system operated by a third-party service supplier are, to a large extent, identical to those of using the services of an umbrella information provider on Prestel: access to expertise and resources, convenience, and economies of scale. The main disadvantage remains the risk of loss of control which may, if this is a significant factor, cause the information owner to operate his own private viewdata service.

The future will see, I am confident, the emergence of the private viewdata system service industry as a significant force competing with both the Post Office systems and hardware suppliers for a share of a multi-billion dollar international market, and their success in this field will reflect their ability to offer a higher level of customer service and flexibility than can be obtained from the Post Office or in-house systems. The emergence of the more flexible packet-switching data networks in the 1980s and the anticipated trend, initiated by the West German Bundespost, to distinguish clearly the roles of the post office as a common carrier of information and the holder and operator of database services, will accelerate the blurring of the existing fairly clear distinctions between those facilities offered by the PTTs and those by the private-sector service industry.

Other services
So far I have confined myself to those aspects of the service industry directly concerned with the design and operation of information or computer systems. There are many other activities which play, and will continue to play, a major role in the rapid development of viewdata technology. Education in the use of viewdata and its promotion through seminars has almost exclusively been provided by the service industry. To the name of Butler Cox and Partners must particularly be added that of Mills and Allen who have, under contract from the BPO, run a series of one-day conferences on the successful design and management of viewdata information systems. As awareness of viewdata technology spreads both within the UK and overseas, and with the availability of private viewdata systems, the demand for informative and useful education and training will grow dramatically.

The emergence of publications to support the development of viewdata has played a significant part in its success. As well as books, there are emerging a growing number of directories designed to assist viewdata users and information owners to make best use of the medium. The leading publishers in the UK are Eastern Counties Newspapers, who publish a comprehensive directory of information on Prestel; the *Financial Times*, with a specialist directory aimed at the business market; and IPC, with a magazine for the residential market and a directory for the travel industry.

162

Consultancy has a definite role to play, in researching new markets for viewdata technology and information systems, in assisting in their evaluation, and in the selection, design, and installation of private viewdata systems. Wherever there is rapid change there will be a strong demand for consultancy services and viewdata technology is going to undergo many changes in the next 20 years.

The computer software industry has already started to have an impact on viewdata. Mention has been made of Langton Information Systems, who pioneered the development of software for automatically and simply converting information held in a conventional computer to a format suitable for viewdata, and who will continue to enhance and develop their system to ensure that it satisfactorily meets the demands of both international markets and private viewdata systems. CAP (Computer Analysts and Programmers) have done much pioneering work on the development of telesoftware, the transfer of programs from a viewdata database to user terminals instead of data. INSAC, the British state-owned software house, has successfully developed software to enable viewdata systems to run on US mini-computers to allow them more successfully to attack the US, and international (including paradoxically the UK) market.

Other software houses are now beginning to become alerted to the UK and overseas market opportunities offered by the advent of viewdata technology, particularly in the field of private viewdata systems, and with the inevitable move towards computer-generated databases the demand for their services will increase rapidly.

Future prospects for the viewdata service industry

As Western industrial society matures, the service industry has emerged as one of the fastest growing. Within Western economies the computer industry is one of the fastest-growing industries—with the computer service industry the fastest-growing segment of that industry. With the advent of cheap microprocessors the information revolution is proclaimed, and viewdata is one of the fastest-growing sectors of the information processing and dissemination industry. I believe history and current trends point to a rapid growth for the viewdata service industry and that the embryonic £1 million industry that it is at present will, over the next decade, grow into a significant industry in its own right, with major export prospects for those organizations with the vision and investment needed to ride with the viewdata storm as it travels across the world.

13. The role of advertising

Alan Jones

Advertising on Prestel: a conceptual problem

The concept of advertising familiar to publishers has an uneasy existence in the new electronic information media. Prestel tends to blur the traditional distinction between editorial and advertising by demanding such a high information content for its advertising that for the viewer the distinction ceases to exist.

As an advertising medium, Prestel had characteristics that sharply distinguish it from other media. It is not like television advertising in the UK where you are hit by a commercial break in the middle of your favourite television programme. It is not like newspaper advertising because an information window of 22 characters down and 40 characters across severely limits any combination of editorial and advertising on the same Prestel page. However, the critical difference from other media is that the initiative to look at information lies with the viewer. Given that advertising information has to be conciously sought by the user, can we talk about advertising at all? The ambiguity about the role of advertising on Prestel is further reinforced by the Post Office common-carrier or open-house policy of allowing anybody to become direct information providers and publishers on Prestel. Organizations whose information, when displayed in other media, carries the label 'advertisement', do not necessarily carry that label on Prestel. The Post Office common-carrier policy does not allow it to distinguish between information providers, and it is the responsibility of information providers to describe the status of their information. The concept and policy of Prestel allow advertisers in other media to by-pass traditional publishing organizations and become information providers and publishers directly themselves on Prestel, with no obligation or reason to describe themselves 'advertisers'.

In practice, however, such companies—let us still call them 'advertisers' for want of a better word—often use the service of the large publishing organizations on Prestel which are described as umbrella publishers. These umbrella publishers offer design and input services, and offer a more cost-effective entry on to Prestel for 'advertisers' than the direct booking of Post Office pages by the advertisers themselves. Fintel, for example, has attracted a major company like American Express to utilize its design and input services. As a leading business information provider, Fintel also provides a context for business advertising on Prestel and has consequently attracted advertisements from companies, banks, actuaries, management consultants, etc. Thus competitive pricing and the creation of a context for advertising have been the umbrella publishers' response to the challenge inherent in the Post Office's common carrier policy which in principle allows advertisers to by-pass publishing houses.

The costs of being a full information provider are set out in Table 13.1.

Table 13.1 Information providers—some direct costs (£) on Prestel UK (1979)

Item	Database size (pages) 100	1 000	5 000	10 000
Entry fee	4 000	4 000	4 000	8 000
Page rent	400	4 000	20 000	40 000
Terminals	2 030	2 030	6 090	10 150
Communications	250	1 000	3 000	6 000
Editor (s)	5 500	5 500	22 000	33 000
Supervisory	2 000	4 000	8 000	10 000
O/hds	3 750	4 750	8 500	11 750
Total	£17 930	£25 280	£71 590	£118 900

These are not the full costs of being an information provider to the UK Prestel service, only *direct costs*. It is therefore not surprising that umbrella organizations, having paid for terminals, editors, etc, can often offer a more cost-effective entry on to Prestel.

Revenue, advertising, or information?

The drive to use Prestel as an advertising medium by publishing organizations has been fuelled by the economics of being an information provider. At the start it was believed that revenue would be generated by page accesses, that is, by a publishing organization charging a price per page for the information it supplied. The figure of £50 was expected to be the average expenditure, excluding telephone charges, of the average viewer per annum. The Post Office forecast of 3 million Prestel sets by 1983 would generate an income of £150 million pounds per annum, half of which would accrue to the Post Office and the other half to information providers. Today UK information providers and publishers, while still concerned with the price of their information, are equally concerned with sub-letting or selling advertising pages on Prestel as a source of revenue. The kind of label attached to sub-let pages tends to reflect the nature of the organizations doing the sub-letting rather than any principle about what does and does not constitute advertising. The position in the UK at the moment, where the different publishing organizations apply different standards to the labelling of their information, is fraught with difficulties. The UK Association of Viewdata Information Providers has drawn up a Code of Conduct for Advertisers and Information Providers on Prestel. This Code of Conduct seeks to ensure self-regulation by information providers and to prevent the legal intervention which abuses of the new medium would invite. The Code of Conduct will also strengthen the position of umbrella publishers, as they will administer the code

165

on behalf not only of themselves but of the Post Office. The Post Office has increasingly provided *de facto* recognition that umbrella publishers are its most effective instrument in minimizing abuse of the Prestel medium. Thus the common-carrier policy is moderated by the requirements of orderly regulation and a code of conduct.

Advertising applications

Prestel during the course of 1979/80 has witnessed a variety of advertising applications. The most significant are as follows:

Adflashes

The advertisement flash or 'adflash' technique consists of drawing attention to an advertisement on another page by flashing characters to induce you to request that page, and has been developed in particular by the *Birmingham Post and Mail,* or Viewtel 202, as it is known on Prestel. The adflash technique attempts to mix editorial and advertising in traditional newspaper fashion. However, the reader/viewer for the first time has to press a button to see advertisements, and little evidence has been produced so far to indicate that someone who is interested in, for example, a news story on President Carter's foreign policy will then press a button to find out what is for sale at the local department store. The efficacy of adflashes is questionable at this stage in the development of Prestel.

Classified advertising

The potential for small advertisements on Prestel is enormous, but the small information window constricts the number of useful classified advertisements that can be displayed at any one time. The Prestel information window allows between one and three classified advertisements per screen, as opposed to the hundreds and thousands that can be scanned by a reader of a newspaper. Prestel does, however, have advantages for classified advertisements because of the live nature of the medium; the advertisements can be deleted as soon as the items are sold or the vacancies filled. Prestel research to date indicates that classified advertising on Prestel is not currently a successful application of the medium. However, these tentative Post Office researches have in no way inhibited one regional newspaper group in the UK, Eastern Counties Newspapers (Eastel), in its plans to feed classified advertisements directly from their typesetting computers on to Prestel.

Product advertising

A discount warehouse, Comet, was one of the first organizations to display lists of goods and updated prices on Prestel. Stores like Curry's and Debenhams have also been active in promoting their products on Prestel to the consumer. However, the product advertising potential of Prestel is restricted by the lack of high-resolution graphics and video images. The technology required for computerized video graphics has been developed in Canada and may eventually be adopted in Britain, allowing for genuine electronic brochures and catalogues with video images in the next few years.

Recruitment advertising

Recruitment advertising on Prestel has attracted both government agencies and commercial organizations. A government agency like Professional and Executive Recruitment has mounted a comprehensive and useful service including a viewdata recruitment service geared towards Prestel's current audience. Commercial organizations like Career-data and Peat Marwick Mitchell are also specializing in graduate and management selection respectively on Prestel. The ability to browse through vacancies of a specialist nature at one's home and office has its attractions, and Prestel will undoubtedly reinforce the development towards a more mobile society. The major problem from the user's point of view with recruitment advertising at present is the lack of any indexing system across information providers at the level of individual jobs. However, despite such problems, recruitment advertising on Prestel has a promising future.

Corporate advertising

Companies have used Prestel as a corporate or financial advertising medium and their advertising services carry the latest press releases, financial results, chairman's statement, etc. While it is too early to discuss the failure or success of corporate advertising on Prestel, one can be confident of companies in ever increasing numbers utilizing Prestel to promote themselves either as information providers or as advertisers.

Sponsorship

This is an area with considerable potential. Sponsorship in the English language does not carry the connotations of advertising and may be a more attractive way of externally subsidizing Prestel information services. Fintel, for example, has foreign banks which sponsor information about their countries and provide useful business and travel information free to the viewer. Sports information is an obvious area of application, with Cadbury Schweppes bringing the viewer the latest test-match score.

The applications described above were developed on a Prestel audience of about 2000 and one should be cautious about drawing too many hard and fast conclusions at this early stage.

The interactive features of Prestel and advertising

The interactive features of Prestel allow the advertising industry to conduct research into advertising as well as to directly market products and services. One of the features of Prestel is that the Post Office computers can be interrogated to ascertain the number of page accesses advertisements have received. This number count allows an access profile of an advertisement to be built up, and on the basis of this advertisements can be redesigned and developed according to the requirements of the market place. This research has a value independent of Prestel in identifying information which is actively sought by the public from the advertiser.

The Post Office also publishes a monthly access table of each information provider. Thus Prestel in effect has a ratings chart in operation. The most popular databases on Prestel (March 1980) were as shown in Table 13.2.

Table 13.2 Top ten information providers to Prestel (March 1980)

Name	Description	Accesses
1. Baric	Computer bureau	141 642
2. Viewtel	Part of newspaper group	86 465
3. Eastel	Part of newspaper group	80 440
4. Family Living	Magazine group	61 566
5. Sealink	Travel organization	51 806
6. Fintel	Owned by publishing interests	51 452
7. Consumers Association	Consumer advice	49 914
8. Mills & Allen	Part of advertising, etc., group	40 950
9. IPC	Newspaper, magazine publishers	37 790
10. British Rail	Train timetables, etc.	37 184

Another feature of Prestel is the response frame. The response frame is a message facility between viewer and the information provider. It is possible to send messages with numeric and with alphanumeric keypads. The method of operation of the numeric response frame is for a user to key a number against the required product or service. The response frame is then sent to the supplier of the product or service requested. With the alphanumeric keypad it is possible to enter into electronic correspondence and operate a system of electronic mail.

The response frame may also incorporate a credit-card number whereby a viewer may buy products through his/her Prestel set. For example, it is possible using your American Express Card to buy bottles of wine from Arambys wine store, towels and a dictionary from Debenhams, and books from Associated Business Press and the International Institute of Communications. Thus Prestel is opening the road to 'teleshopping'.

The frame count and rating facilities on Prestel will become essential tools for the media planner and will form in future the basis of ineter-media comparisons. The access profiles will shape the future of Prestel advertising in accordance with the market place, and the response frames and credit facilities will provide additional measures of Prestel success and inter-media comparison.

Prestel and other media

The future of Prestel lies not only in its acceptance by the public but also in its acceptance by and integration with other media. The interactive features of Prestel are natural adjuncts to existing media which are on the whole passive and unresponsive. How is this integration proceeding in the UK? We are beginning to see organizations using their viewdata numbers in newspaper advertisements. These advertisements refer people to Prestel for more detailed information on their products and services. Prestel services run by newspaper organizations like Fintel, Eastel, Birmingham Post and Mail, are designed to be complementary to the printed edition of their newspapers and not competitive.

Oracle, the independent television teletext service, is currently publishing advertisements. Fintel has placed American Express advertisements on Oracle which refer viewers to more in-depth information about American Express on Prestel. Unlike in France, there is very little interest in teletext–viewdata collaboration in the UK. Teletext also offers less commercial opportunities in the UK than the Antiope service licensed by TDF in France. There it is possible to operate a number of commercial teletext services, as the French intend devoting a whole television channel to teletext.

Television commercials in 1980 have begun to promote Prestel as a media option, and Prestel will now have to enter into competition for the use of the television set against competing functions. Possible levels of ownership in UK homes of the different television-based functions in 1985 have been estimated by J. Walter Thompson and ITT as follows:

Television games	14 per cent
Video cassette recorders	10 per cent
Video disc players	5 per cent
Teletext	2–10 per cent
Prestel	1–5 per cent

(Admap, April 1979)

The lower estimates for teletext and Prestel indicate doubt on whether receivers will enter mass production: the higher figures assume they will. The Prestel forecast of between 1 and 5 per cent is the most pessimistic forecast published in the UK for 1985, but still leaves Prestel with an audience of 200 000–1 000 000 by 1985, making it a significant medium against competing functions.

There is to be an additional television channel in the UK which will result in increased programming. It is also likely that satellite television stations will appear in Europe, beaming programmes to the UK in the late 1980s and early 1990s. Contenders for satellite television stations in Europe are Tele-Luxembourg and possibly some United States television satellite companies. What is certain about the future is the availability of more television channels.

The video cassette market is forecast to grow, with 10 per cent of UK homes owning video equipment by 1985. It is important for Prestel that a cross-over technology develops whereby video and data technologies can be combined. The video disc offers exciting possibilities for combining off-line video displays of, say, holidays, with on-line Prestel booking facilities in hotels and travel offices. American Express is currently experimenting with combining video discs and Prestel. The video-disc market may pose a competitive threat to Prestel-type information services which are based on historical data, as entire encyclopedias

169

can be stored on video disc. The video-disc market and the Prestel market are likely to be the same size during the mid-1980s, according to J. Walter Thompson and ITT, and possibly offer profitable combinations.

The continued growth of remote-control keypads can be expected. This makes it easier to choose between television-based services and may well threaten the effectiveness of commercial breaks which are increasingly being used to scan other channels and, in future, other media options.

All this competition for the use of the television set will tend to generate demand for extra television sets and it is estimated that by 1985, 25 per cent of all UK homes will have two television sets and by 1990 we will have a television set each, and television-based services will become more individual than social.

The television audience meters currently in operation in ten countries record the station to which audiences are tuned, but are inadequate for the audience-measurement requirements of the television set of the future, with teletext, viewdata, videocassettes, video discs, etc. Viewdata and Prestel, however, by linking the telephone line to the television set, open up the possibility of new interactive television-audience meters and a wealth of inter-media statistical data. One of the problems of a new medium like Prestel is that it does not initially fit into the cost-per-1000 formula of the advertising world or the minute-by-minute ratings of television.

Chris Powell, the Managing Director of Boase Massimi Pollitt Univas, has offered some tentative suggestions to his advertising colleagues on the cost effectiveness of Prestel. His calculations suggest £8/£9 as a cost per thousand with 250 000 sets in the market place. This is far from being a discouraging figure given that we are talking about a much more targeted audience, more likely and able to convert advertising to sale there and then.

The question of inter-media comparisons is at the top of any agenda that seeks to encourage any wholesale movement of support by the advertising world for Prestel and the viewdata medium. It could well be that it is the support of the advertising community that determines whether Prestel is just another interesting computerized information service or a medium for the millions that many of us wish it to become.

Conclusion

The UK experience strengthens the view that Prestel is a promotional medium and that major advertisers will utilize the publishing services of umbrella organizations to promote their products and services. The umbrella publishing organizations are well placed to integrate advertisers and advertising agencies into their Prestel activities as they already have established business relationships and commission structures with the advertising community. It is noteworthy that many of the top publishing organizations on Prestel are newspaper or magazine publishers. Newspapers which are computer typeset also have the possibility of directly feeding their classified advertisements into the Prestel service and generating additional revenue from existing advertisements.

The economics of Prestel are analogous to newspaper economics with Prestel in the UK at the stage of viewers having to pay the full economic price for their service. The exterior subsidy which moves newspapers into the market place of millions is also required for Prestel, and the advertising community is the means for that change of Prestel into the medium for the millions.

To secure the wholesale involvement of the advertising community, research into inter-media comparisons is required and new television- and telephone-based meters for audience measurement are a necessity. In the UK, Prestel can take a lesson from commercial radio, which failed for years to carry out the necessary research to aid the presentation of its advertising case, and as a consequence failed to maximize its advertising opportunities until it belatedly undertook the expensive research necessary for the effective presentation of its case.

14. A code of conduct

Richard Hooper

The otherwise rather boring subject of codes of practice raises in the case of Prestel, the British Post Office's viewdata service, some rather interesting issues, issues that take us to the essence of Prestel as a new communications and publishing medium. Why, in the fist place, is there a need for a code? The short, if rather facetious answer relates to the climate of opinion in most Western mixed economies that 'if it moves, it should be regulated'. The rise of consumerism in the West has led to a rise in codes of both a statutory and non-statutory variety. But in the case of Prestel, the arguments for a code are rather more self-evident, though the content and implementation of the code are certainly not.

Because of the BPO's so-called common-carrier policy towards Prestel information providers, the BPO cannot and does not wish to exert direct control over the content of the pages. If the content is within the law and not grossly offensive (a Post Office addition which has no strict statutory meaning), then as a general rule that content is not a matter for the Post Office, as Prestel system operator, to concern itself with. A quick glance at Prestel will be enough to show that all sorts of practices could emerge which were not illegal, not grossly offensive, but which were undoubtedly a 'rip-off' of the Prestel customer. For example, the Prestel user pays a price for a page as soon as that page is keyed. Thus there is an opportunity for the information provider to route the unsuspecting user from the current page priced at say, 1p, to another page which will be priced at 50p. A code is clearly needed to ensure that price changes are signalled *in advance* of the actual keying. A further reason for having a code of practice concerns the notion of the public good. In essence, Prestel will be perceived by the user as only as good as its worst pages. The user will not make subtle differentiations between different information providers, not initially anyway. Bad pages will drag down good pages. Good information providers will suffer from less worthy colleagues. Thus it is in the interest of all within the Prestel community that a general standard of behaviour should be agreed and adopted.

If there is agreement as to the need for a code of practice, who should produce it and provide the necessary leadership? In the case of Prestel, only two agencies seem relevant—the Post Office and the information providers. There has been no interest expressed within the government so far concerning a statutory code for Prestel. Traditionally in the UK, codes have been non-statutory and self-regulatory, for example, that administered by the Press Council. Given their common-carrier stance, the Post Office could not and indeed is not interested in creating and administering a code of practice. However, the public-service contracts which information providers are signing with the Post Office require the information providers to adhere to certain 'standards'. These could be construed as having a code of practice as their objective. The Post Office is obviously concerned about behaviour by information providers that is within the law, but is potentially harmful to Prestel's development as the modern-day equivalent to Gutenberg's printing press. If Prestel was to get a code of practice in the short term, it became clear in early 1978 that only the information providers were in a position to take the initiative. Some information providers felt that the creation of a self-regulating, voluntary code without further delay would be a pre-emptive strike against other regulatory or government bodies getting involved and slowing up developments, as has been the case in other European countries. From the beginning, the Post Office was very supportive of the idea of a self-regulating code produced by the information providers.

At the time when a code of practice was first discussed in Britain, at the start of the Prestel test service, it was fortunate that the information providers were forming their own trade association: The Association of Viewdata Information Providers (AVIP). AVIP considered that a code was high on the list of their early priorities. As a result, the council of AVIP set up a committee to establish a code of practice. This committee worked on the subject throughout most of 1979. Drafts were sent out to information providers, and to other interested parties such as the Advertising Standards Authority (ASA). The final version was approved by the council at the end of 1979. The main difficulties which the committee found itself faced with, concerned the following issues:

- definition of advertising on Prestel
- one code or many codes?
- Prestel or viewdata?
- record-keeping requirements
- sanctions and scope of the code.

Definition of advertising on Prestel

A code of practice on Prestel depends crucially on defining what is information and what is advertising. Prestel as a medium tends to blur the distinction. For example, in print the advertising copy is differentiated from editorial copy by the use of different typography and layout. The Prestel television receiver has only one character set, thus making typographical distinctions very difficult. Because of the limitation of typography and graphics on Prestel, there is a general tendency for all pages to look alike, irrespective of content. A further problem is caused by the fact that each Prestel page carries in the top left of screen the name of the information provider. This fact ensures that each page promotes, directly or indirectly, the name of the information provider. Where the name of the information provider is well known in publishing circles, for example IPC or Guinness Superlatives, then there is automatic promotion of IPC and Guinness publications.

There is a school of thought which asserts that, because of the medium's characteristic blurring of advertising and information, there should be no attempt to discriminate between the two. This view seems to be rather naïve. The existing codes of practice in the UK, notably from the ASA governing print advertisements, and the Independent Broadcasting Authority (IBA) on commercial TV advertising, concern themselves almost exclusively with what can and cannot be advertised in the respective media. In the medical area, for example, there is an enormous amount of statutory rules and regulations which underpin the codes. If Prestel's code were to ignore the control of advertising, it would be asking for immediate trouble from many quarters. But there is a more commonsense reason for having to make the distinction. If the viewer sees a page on the screen about the Hotel Norwich, for example, then he or she is entitled to know whether the description of the hotel was written and funded by the Hotel, or by an independent hotel guide hired by the information provider Eastel. However, even this simple example becomes complicated on Prestel. If Hotel Norwich decides to register directly with the Post Office as a Prestel information provider, rather than take paid space on Eastel's page, then the traditional distinction between the media owner (for example, the newspaper or television station) and the advertiser (in this case, the hotel) is lost altogether. The advertiser becomes the media owner directly. The advertiser may then assert that it is no more necessary to put the word 'advertisement' on all its pages, than it is to put the word on any print promotional brochures it produces.

It should be clear by now that this problem of definition is very difficult. After many attempts, the AVIP committee came up with a defination of advertising on Prestel as follows:

'On viewdata systems, an advertisement will be deemed as:

(a) any frame or part frame whose prime purpose is to promote the sale of goods and/or services other than the actual information stored on viewdata itself or to promote the reputation or image of a company or organization;

and/or

(b) any frame or part frame which carries an announcement to promote the sale of goods and/or services other than the actual information itself or to promote the reputation or image of a company or organization, for which a consideration has been paid to the provider of that frame. However, a consideration paid by a sub-information provider to a main information provider for frame space does not necessarily constitute advertising.

Where the source for information displayed on viewdata is identified, this identification will not necessarily be construed as advertising.'

Where the content constitutes an advertisement according to this definition, that content on Prestel must be clearly labelled 'advertisement' or 'advert'. (The latter option was given because of the high premium on space on a Prestel page—only 22 lines of 40 characters.)

Finally it is worth noting that the section of the code on information is about one twenty-fifth the size of the section on advertising.

One code or many codes ?

The debate here concerned the relationship between the new AVIP code and the existing codes such as those from the ASA and IBA. One group of information providers felt that the new code should cross-refer to the existing codes rather than try and be yet another code. Another group believed, equally strongly, that Prestel was a new medium in its own right which therefore deserved its own code. However, any such new code would draw heavily upon existing codes *where relevant*. This latter view won the day. The ASA code became a central source of ideas and phrasing for the AVIP Prestel code, but was open to modification. An example of a modification concerned the rules about corporate advertising. Business information providers like Fintel felt that the ASA wording was inappropriate to Prestel. Another example concerned the IBA code, on the matter of cigarette advertising. Cigarette advertising is illegal on commercial television. The AVIP code felt that no such constraint was justified. Cigarette advertising is allowed in print, therefore it should be allowed on Prestel which is not a broadcast system.

Prestel or viewdata?

Should the code be about Prestel or viewdata? It was decided that the code should concern 'all frames for which AVIP members are responsible, which are in any way intended to be accessible by the general public'. This therefore covers any public viewdata systems, but not private viewdata systems. The Stock Exchange, for example, felt it was unreasonable for the code to cover any private viewdata system that the Stock Exchange might develop for its stockbroker members. AVIP felt that Prestel was not likely to be the only public version of the viewdata concept and for that reason had not called itself APIP (Association of Prestel Information Providers). However, pragmatically, the code is very much based on Prestel as it exists now, while the actual wording is more generalized. The term 'Post Office' is largely replaced by 'viewdata system operator', for example.

However, this widening of the code beyond Prestel to viewdata will also lead to difficulties of definition. What is a public viewdata system? If existing publicly available on-line databases take on some of viewdata's characteristics (for example, the terminal/TV set), would those databases then come within the code of practice? What started out as a code of practice for Prestel specifically, may actually develop into a code for the emerging public computer systems, those that are in the jargon 'externally available'.

Record-keeping requirements

In many codes of practice, there is a voluntary or statutory requirement for the media owner to keep records of the editorial and advertising content. Newspaper companies, for example, normally keep back copies for as long as seven years. It seemed natural that Prestel should expect similar behaviour, both from information providers and advertisers. However, AVIP came to a pragmatic decision that the medium of viewdata, which is based on 'soft' copy that is quickly erasable, should not necessarily follow print habits, where 'hard' copy is part of the production process. The requirement on information providers to keep records of all their pages, plus all their changes to those pages, would quickly become horrendous in scope and cost. Legal actions arising out of complaints concerning Prestel pages will have to draw on rules of evidence that are used in actions concerning what was or what was not said in, for example, a telephone conversation. It would be true to say that none of the AVIP committee felt that the record-keeping problem had been satisfactorily resolved—but instead, a defensible interim policy had been adopted.

Sanctions and scope

The final headache for the AVIP committee concerned the sanctions and scope of the code. The question of scope was more easily answered, though many were not satisfied with the answer. Because AVIP had decided to take the lead on the code, any code produced could only refer to AVIP members. Constitutionally, it is obvious that an AVIP code could not be used to discipline non-AVIP members. For example, who would do the disciplining? As it happens, AVIP members account for some 80 per cent of the Prestel database, but the problem still remains—what about pages not covered by the code? AVIP cannot solve this issue, and it remains to be seen if government will at some point require the Post Office to do so, or set up some body specifically with this task. It is worth noting that even the ASA's code of practice does not have 100 per cent power over printed publications.

The problem of the sanctions behind the code—the teeth—have also not been resolved to everyone's satisfaction. Given the scope of the code, AVIP members, and its overall philosophy, voluntary and self-regulating, the only real sanction is expulsion from membership of AVIP. Not a very fearsome sanction, since it does not mean expulsion from Prestel itself. The Post Office has put forward the idea of a logo both on-screen and in the directories which would show that the information provider subscribed to the code, rather like the seal of approval from Good Housekeeping. This logo would be removed if the information provider was in breach. As with self-regulating codes, the sanctions are not very sharp in practice.

The sanctions will be the responsibility of a Complaints Committee to be set up by AVIP. This Committee will have a 4 : 3 majority of independent members over information providers. How the independent members will be elected is not clear. No existing agencies seem either able or willing to help. When and if a Prestel users' association is formed, that would be the natural body. In the judgement of the AVIP committee on the code of practice, most complaints will not reach the Complaints Committee because they will be resolved 'out of court'.

Conclusions

The AVIP code is written with change in mind. As the preamble states: 'The Council of AVIP accepts the need to update this Code of Practice continuously as the medium develops, and generally in advance of anything which could be achieved by legislation.'

As public viewdata systems like Prestel take off and become mass media, then the spotlight of public scrutiny will be trained on the editorial and advertising practices of all organizations publishing via Prestel. Then, and only then, will politicians become interested in regulation and legislation. Record-keeping will, in my opinion, become a major topic of debate, as will the scope of the code. Greater controls over advertising will almost certainly be introduced. Prestel will know that it has really arrived as a new medium in its own right, when the law is changed to ban smoking advertisements on Prestel pages!

Copies of the AVIP Code of Conduct can be obtained from AVIP, Minster House, 12, Arthur Street, London EC4R 9AX at a price of £5 per copy.

Part Five

International survey

```
FINTEL                    2481a            Op

Prestel         Viewdata        Transpac
Viditel         Videotex        Didon
Telset          Teletext        Titan
Telidon         Text-TV         Green
                                    Thumb
Bildschirmtext  Infotext
Teletel         Ceefax          Bildschirm
                                    -zeitung
Captain         Oracle
Vista           Antiope         Mercury
Viewtron        Qube            Euronet
Ida             Tictac          Diane
```

Editor's introduction
Faces of the polymorph

The international development of viewdata systems is marked by great variety of approach and philosophy, as well as by substantial variations in technology, which nevertheless overlay a basic family unity. The following chapters explore this variety of national approach, rather than necessarily giving a complete factual description of each national system. But several major themes can be discerned in the pattern of international events.

1. Should viewdata be a self-contained information system, or should it be a 'gateway' to data held elsewhere? In other words, there is a substantial difference of approach between, for example, the British Post Office with its plans for a large network of computer centres covering Great Britain, all of them holding more or less original data in the viewdata format and acting as the basic 'warehouse' of the viewdata information content; and the French and Germans, both of whom with varying degrees of emphasis see viewdata as the gateway between the user-customer and the holder of the information that the user wants. In this latter situation, the viewdata computer or computers merely act as traffic managers in the middle, receiving a request for information, relaying it to the computer centre at the information provider's own premises, extracting the required information, reformatting it into a viewdata display, and transmitting it back to the original enquirer.

   This approach has the advantage that it avoids the necessity of building up an expensive network of PTT computer centres, and avoids the networking and updating problem (keeping the data up to date, and correct, and the same, on all computers at the same time) that is inherent in the British approach. On the other hand, it represents in itself a complex computing and communications problem which has yet to be tried and proved successful in practice. However, in a society that already holds large amounts of data in computer-readable form, such as is increasingly the case in Germany, the Netherlands, and (presumably) France, it must be a rational approach. If nothing else, it avoids the expensive business of re-creating data, manually or by bulk update, from one system to another. What can be said about the British approach is that it works, and that for Britain at any rate it may be the only one that can work for some years to come.

2. There is equal variation in the importance attached to graphics. Prestel has a real but limited graphics capability: real in that it can enhance the attractiveness and presentation of the data, making use of the fact that viewdata is intrinsically a colour medium; limited in the sense that the geometric style of graphics creation and the lack of finesse in the basic

graphic building blocks make it hard to create all but the crudest images. Graphics are surely a secondary feature.

But the French have enhanced the graphics capability in their own viewdata design, and the Canadians have improved it even further, to the point where the graphics capability is really at the heart of the medium, not a secondary feature but a, perhaps the, primary feature. So how important are graphics in 'selling' viewdata to users? Perhaps an even more important question is, what can a viewdata system with first-class graphics do that one without them cannot? Do good graphics—for maps, diagrams, pictures, games, amusement, advertising—fundamentally transform the medium and its capabilities? Or are they not worth the substantial extra cost that, at this stage, is undoubtedly involved? Is it better to get a cheap and cheerful version like Prestel out into the market place sooner, or a less cheap, more sophisticated version like, say, Canada's Telidon, into the market later?

3. Is viewdata a residential or a business system? Or both? And are the time scales for business and residential penetration different, and if so, which comes first? The British approach was to say that it was a mass residential medium, depending as it does on the telephone and the TV set, both mass-market phenomena; coupled later with a growing awareness as time passed that there was greater and growing acceptance of viewdata in the business market, so that business might become the initial take-off area, leading to wide domestic use later. At the time of the full public launch of Prestel in spring 1980, this divergence had not been properly resolved—perhaps it cannot be. But both the French and the Germans are determined that their market trials shall be fully residential in nature, despite the evidence from Britain that business is where first acceptance comes. Conversely, in countries like Hong Kong viewdata can hardly be anything but a medium of business communications, if it is to succeed at all.

4. The relationship between viewdata and traditional publishing media like newspapers shows marked contrasts between countries. In the UK, some newspapers have taken viewdata seriously, the majority have not, or not yet: certainly, there has been no collective attitude or collective action, and decisions have been left to individual publishers to determine what advantage, or disadvantage, there is in participation in viewdata. But in the Netherlands there is a concerted attempt to place the whole newspaper industry, title by title, on the Dutch Viditel viewdata system, as a collective newspaper information service with a unified structure and a separate slot inside it for each newspaper. In West Germany also, as in the Netherlands but unlike Great Britain, there has been a strong tendency to reproduce the titles of newspapers and magazines on viewdata, to emphasize the interdependence of the two media and (presumably) defend the position of the newspapers. The idea that viewdata is some sort of electronic version or electronic 'clone' of the newspaper is a controversial one, deeply embedded in many publishers' minds, but one that is rejected by many, including this author. It is discussed fully elsewhere.

5. Is viewdata an information medium, or will it be successful when it is more active, more interactive than that, when the capability of the user to answer back to the computer, to order goods and services or send messages to where he will, is fully developed? Or is it even more extreme than that? Several of the Canadian experiments with viewdata see it strictly as a functional medium for monitoring burglar alarms, regulating central heating, reading electricity meters, conveying fire alarms, and otherwise coping with the routine management of problems of an affluent North American household. Provision of information will, on this theory, be secondary to, and ride on the back of, these functional applications.

6. There is strong competition developing in world markets for sale of viewdata systems, especially between the British and the French, who have formed some sort of technical/marketing alliance with the Canadians. The Japanese, while interested in European and US developments, have clearly set their sights on capturing the market for viewdata systems in those countries, like Japan itself, that do not use the Roman alphabet. But for the Roman-alphabet world, competition may get fiercer.

   The British have been marketing overseas in at least three ways. Firstly, the British Post Office has been selling the Prestel system direct, with an active sales team, and has made sales to the Netherlands, West Germany, Switzerland, and Hong Kong. Secondly, it has licensed the software company Aregon, formerly known as Insac Viewdata and owned by the National Enterprise Board of Britain, to rewrite the Prestel software for other (e.g., DEC) computers and sell it abroad, notably in the USA. Thirdly, it has commissioned a British computer consultancy, Logica, to set up and manage a market trial for Prestel International, which as the name implies is a service operated from London, on computers located in that city, but marketed internationally to multinational companies to meet at least some of their needs for internal information transfer and information gathering.

   But the French have been very active in promoting their rival system in competition with the British efforts, despite not yet having in France an operational viewdata system. There is every sign that the French intend to give the British a run for their money.

7. France also illustrates a profound difference of political approach to viewdata development. The British view has been, fundamentally, that of letting market forces determine the future of Prestel, and that includes a simple commercial decision by the British Post Office to develop the system at all. There is little if any evidence so far of direct government intervention to speed up or initiate viewdata progress (which is not to say that various government departments have not supported Prestel by seeking to make use of it for their individual information purposes—the Central Office of Information has been active and often successful in persuading them to do so).

   The French government has, however, taken a conscious decision, on national policy grounds, to develop the information handling and

183

distribution capacity of French society by a variety of means, such as improved telecommunications and new services that include viewdata. In particular, the decisions (yet to be proved in practice) to develop the electronic telephone directory for thousands and ultimately millions of French households, could be a bold and imaginative move initiated by central government to establish viewdata as a normal means of mass communication.

8. There is marked contrast between this quasi-political keenness to develop viewdata, with ambitious plans to penetrate the majority of households, and the hesitation and doubts that characterize US attitudes to viewdata. In a free-enterprise society that already has a mass of communications facilities, such as cable, satellites, data networks, as well as a mass of TV and radio, there is as yet no substantial commitment to viewdata, despite the efforts of GTE, Aregon, and Knight-Ridder, described later. This must be a cause for concern, both because failure by the Americans to adopt viewdata would damage its reputation in the world at large (though not necessarily fatally) and because support by the USA, with its manufacturing facilities and multinational communications companies, would be a powerful boost to international developments. There may be a strong element of 'not invented here' in the American attitude. But it may equally reflect the quite valid view that there is nothing special about viewdata: it is one of a number of competing computer-assisted communications media that has to justify itself in such a competitive environment, and may indeed within a few years (as some experts predict) lose any separate identity and merge back into the general computer environment from which it emerged. If the efforts being made in Britain, the Netherlands, West Germany, and France are positive plus points for viewdata at this stage of its development, the position in the USA is definitely a minus that must be put into the calculation.

The survey of countries in this section is not totally exhaustive, although the main countries are covered. To them there must be added, as countries with at least some viewdata activity, Switzerland, Hong Kong, Spain, Belgium, Scandinavia, and New Zealand. Switzerland has bought the Prestel system, but has yet to develop a full market trial. Hong Kong has also bought the Prestel system, and is working out its (predominantly business) uses for it in a territory where something like 98 per cent of the population is Chinese, distances are small, and profit is the god. Spain has devised its own viewdata system, but until 1980 had not made much effort to put it on show, even for demonstration purposes. Belgium also has a prototype system, with little yet to be seen in public. The Scandinavian countries (apart from Finland, separately described) has had seemingly endless committees of enquiry into viewdata, producing reports for official or industrial consumption, relating in particular to the question of advertising, effects on the press, and privacy. What will emerge from this mass of committees remains to be seen. New Zealand will have a private viewdata system bought by a private operator, from the UK.

These are the countries at present standing on the periphery of viewdata developments, with many other countries watching and waiting. Viewdata could in theory come to every country with a reasonably developed telephone system, and there is therefore potentially a huge market for exporters of viewdata hardware, software, services, and (not least) information held in viewdata format. Whether that mass international market develops will to some extent be governed by the attitude of influential countries like the USA: for what uses it develops, will to some extent be governed by rival theories being developed in Britain, Canada, Germany, and France: and who secures those markets, in terms of trade, will be governed by technical and price competition between Britain, France, perhaps Holland in the future, and no doubt other producer countries as well. Prestel in the UK has a head start, but will find itself in an increasingly competitive environment.

15. West Germany

Dietrich Ratzke

State of development

The development and introduction of new forms of text transmission using TV sets has been neglected for many years in West Germany. While the UK and France developed their market trial systems, West Germany was still disputing the question of principle. First of all they argued about who would use this new communications system and thus who should provide the information. The answer to these questions was dragged out for years on political grounds and even today has still not been finalized. The development of communications technology was hindered, and the communications industry endangered because each 'Land' or region of West Germany is responsible for its own cultural affairs; cultural autonomy in West Germany includes, among other things, television. From the beginning the broadcasting organizations attached to the Land governments claimed that the new forms of text transmission would be their responsibility as in their view, videotex in either form (text information carried in the spare lines of the TV picture or text sent by the telephone) was broadcasting. The existing law is in principle on their side, stating that broadcasting, in accordance with the 'state agreement of the Land governments on broadcasting tariffs' of 4 December 1979, is defined as 'the function and distribution to the general public of display of all kinds in words, sound, and pictures, utilizing electrical oscillation, either without connecting wires or by means of cable'.

As some wit observed, from this narrow definition thunder and lightning would be 'broadcasting' and so should only be caused by West German 'Land' governments. There was actually a court decision over a lottery-ticket salesman at a fair who used the help of a public loudspeaker to announce the sale of tickets; this was declared 'broadcasting'. Because the Lander governments with the help of this law are so jealously guarding their own cultural and television autonomy, they have prevented private enterprise from offering the new text systems (and also, naturally, cable television and other new media).

The situation only changed when the West German Bundespost in 1976 made known its intention to try out the British viewdata system modified as Bildschirmtext in West Germany. Some of the Lander governments declared themselves ready to allow private information providers during a Bildschirmtext trial. The broadcast videotex system, however, remains closed to private suppliers.

Another thing had delayed the test and introduction of Bildschirmtext. In February 1974, the West German government had set up a commission which, with large financial and scientific resources, would ascertain the necessity and

186

requirement for new information and communication media. This Commission for the Development of Technical Communications System (KTK) submitted after almost two years' work a detailed report in December 1975, which, however, contained only very vague points about the scope of Bildschirmtext, and gave no useful recommendations to the equipment manufacturers and potential information providers. They confirmed that the new forms of telecommunication would be possible by connecting to home television sets through the existing dial-up telephone network. 'The information content can be text or graphics. The subscriber is presented with a multiple choice which leads to the information which he requires. The technical assumptions and economics need careful observation and development' (*Telecommunications Report*, p. 104). From this came the equally unsatisfactory 'recommendation': 'The development of new equipment for reception and reproduction of text and graphics needs intensive research' (*Telecommunications Report*, p. 92). The other general recommendation made by the KTK to install in the beginning broad-band research projects (with cable text) has still not been realized as a result of the politicians disagreeing about the participants and the financial questions.

The SPD (the ruling Social Democratic Party) feared especially that through private participation in the new medium the public-service nature of radio and television would be changed and an 'American state of affairs' would prevail in West German television. The conservative opposition party, CDU, however, advocated the private sector approach. With this conflict, the pilot projects for Bildschirmtext and videotext remained on ice for years.

Bildschirmtext trial Düsseldorf/Berlin
The trial sample
During this political infighting over the medium, the German Post Office seized the initiative; they made it known that from 1 June 1980 in Düsseldorf/Neuss and also Berlin, they would begin a controlled field trial with 2000 private householders and 1000 commercial users. Instead of providing the home users with special sets for the brief duration of the trial, which would have been too much bother, another way was found. In Düsseldorf in the middle of November 1979 they sent a mailing shot to 500 000 families to locate families who were planning to buy a new TV set. Those interested could apply to the Bundespost if they wanted to participate in the trial. They could buy the TV set of their choice from the dealer and would then be supplied with a videotex and Bildschirmtext decoder at the expense of the German Post Office and Industry. The period before Christmas was deliberately chosen for this action as experience showed that around 50 per cent of TV sets sold in a year are acquired in that period. From 500 000 households mailed there were 2800 replies, and from these 1500 participants were selected. As these interested householders belonged, as a rule, to the higher-income classes, 500 were withheld and reserved to provide the necessary sociological weighting for the sample. These additional participants were approached in a limited further action. The 1000 commercial participants were recruited through proposals from the information providers.

187

In Berlin the recruitment of the participants was tackled in another way. During the Berlin broadcasting exhibition in August 1979, cards were given out to those interested in participating. Of these, 1600 were returned by those seriously interested. The remaining 400 places were reserved for allocation on the same demographic grounds as in Düsseldorf.

The Düsseldorf participants can call up information at 23 pfenigs for 8 minutes in the daytime and 12 minutes at night. In Berlin there is no time charge; thus, for 23 pfenigs, one can call up the information for an unlimited time. In addition to the connect charge, there is a charge to the caller for each accessed, which can be fixed by the information provider between 0 and 99 pfenigs.

At the end of 1979—during a pre-trial period—40 000 pages were entered, and by mid-1980 in Berlin and Düsseldorf 60 000 pages were available in each centre. The Post Office system supports 208 simultaneous telephone connections in Berlin and Düsseldorf. A free Bildschirmtext periodical is to accompany the trial. This periodical also contains the official list of information providers. The middle of 1981 has been fixed for a decision about the continuation of the trial. By the end of 1982 the full system should be introduced. In the following year 60 per cent of all telephone users will be able to connect to the system and in the second an third years this will rise to around 85 per cent.

Why Düsseldorf as a test region?

The primary objective of the Bildschirmtext trial is to test the readiness of the private householder to connect to the Bildschirmtext system. To obtain a demographically representative sample population, one must have an urban connurbation which contains both a city centre and satellite towns. The different regional structures must tie up with one another so that conclusions can be drawn about the reactions of the entire population of West Germany. In selecting a suitable test region it was also important to try to predict the future economic and social development in this test region, particularly in respect of the service industries. Naturally the demographic characteristics of the population in the test area ought to be approximately representative of the whole of the country. Not least, an area must be found in which there is sufficient existing media density. Without enough telephone numbers and TV sets, the basis for the field trial would be removed. Above all, the test area must be exactly delineated in order not to falsify the scientifically measured results. For Bildschirmtext to undergo commercial application trials that produce lessons, the economic structure in the planned test area must be acceptable. Professor Heiner Treinen held that the decisive factors in favour of the Düsseldorf/Neuss region which was decided upon by the Scientific Advisory Commission of the West German Post Office were the following:

Düsseldorf and Neuss present a double city centre which, together with the suburbs, have the structure of a medium sized municipality, so that in the test area you have the commuter phenomenon and the orientation towards non-local supplies of information (for instance, mail order catalogues). The area mentioned contains an interesting variability of regions of high and low density population. City centre and suburbs are, because of the social and economic structure, inter-related. Both towns are surrounded by other town regions, so that a fairly clear delineation can be made. The socio-demographical structure of this area does not contain any marked deviations (for example, in terms of sex, age, net income) from other comparable regions. The tertiary sector, characteristic of the trend towards the future 'information' society, has a relative dominance. The economic structure is desirably mixed (for instance, mixed individual trade structure with varying density of consumer markets). In favour of the choice of the city of Düsseldorf was its central function for the supply of media to the suburbs of other towns. Also of importance for the trial is the high density of communication per household in terms of daily newspapers. Also of significance for the choice was the fact that Düsseldorf appeared to be central for potential information suppliers, so that, at the time of the field trial, Düsseldorf can reckon on a varied supply of information. Its high density of telephones and TVs, in relation to other regions, also speaks in favour of this choice. The higher level of education in Düsseldorf and its suburbs corresponds to the expected structure for Bildschirmtext. In addition to this it should be borne in mind that because of the varied landscape—in terms of newspapers and broadcasting—we can reckon on this trial region producing the widest possible public response.

Similar, if not quite so obvious, demographic advantages are to be found in the Berlin region. Who is providing the information in the Düsseldorf trial? At the beginning of 1980, 765 information providers had expressed interest in taking part in the trial, 371 had signed an agreement with the German Post Office, and 212 had begun the trial. At the forefront of the information providers, who at the end of 1979 had entered the lists, were those with general business interests from 97 firms:

Business/Services	26
Banks/Financial institutions	21
Associations	13
Transport/Tourism	12
Insurance companies	11
Industry	10
Consultancy firms/Advertising agencies	2
Computer centre	1
Consumer council	1

The second strongest group of providers comprised the media companies:

Journals and book publishers	32
Newspaper publishers	27
Broadcasting institutions	5
News agencies	3
Others	2

The remainder were:

Manufacturers of Bildschirmtext sets and components	18

Scientific institutions   13
Public administration   6
Churches and social establishments   3

From the so-called computer fraternity, that is, those firms with their own computer installations, at the moment only the mail-order firm Quelle and Otto are active.

The cost to the information providers for the duration of the field trial is 5 DM per month for the modem; and, for the full trial, the connection costs. For the time after the end of the trial, the Post Office plans an entrance fee for every information provider, the level of which has not yet been fixed, but which will be comparable to that for the viewdata entry fee in Britain.

The scientific research

The West German Post Office and the government of North Rhine Westphalia have divided the research between them. The Post Office scientific research will be carried out by Professor Langenbucher of Munich, and others. A pilot project in 1978/79 included opinion polls testing information retention, the perceived shortage of necessary information, the search skills and search strategies, contact with the authorities, and knowledge of Bildschirmtext, through a representative random sample in Düsseldorf/Neuss and independently of the purpose of the Bildschirmtext trial. It tested above all the readiness of the people to pay for new media, and established the market potential.

From the result of these questionaires, a model of the 'typical videotex user' could be deduced. He is largely a member of the so-called upper classes (classes 1 and 2 out of 3 social classes), he earns over 3000 DM a month, he works in an office, he is between 25 and 30 years old, he has a far wider than average range of communication at his disposal, his expectations are far higher than average in dealing with technical communication, both in content and form. Of this ideal type, 2700 to 3600 of those questioned intend to purchase a new television within the next twelve months. Thus the random-sample technique envisaged by the Bundespost with 2000 participants was possible. The test research gave no cause for euphoria in terms of public acceptance of videotex; it did, however, reveal worthwile perspectives in some areas. According to this pre-research, there will not be a rapid penetration of the market, as there was when television was introduced. Furthermore, due to the disproportionate interest shown by the upper social classes towards new information media, a further, growing gulf is expected to open up, in terms of knowledge, between the upper and lower social classes.

Further over-saturation of information by videotex is not to be expected, as the recipient is in a position actively to influence the recall (request for information). During the trial the scientists around Professor Langenbucher will use various different methods of research:

1.   Panel/opinion research, which provides for an enquiry at the beginning of the field trial as well as a concluding questionaire one year later. During the

entire duration of the field trial, regular telephone questionaires will be conducted, and the diary records of the information users will be evaluated.
2. Reports from the videotex headquarters, as far as this evaluation does not contravene the law e.g., the Protection of Information Act.
3. Sub-random samples in the form of intensive interviews and group discussions.

In order to aviod 'over-researching', the suppliers of television sets in the trial are not permitted to conduct any surveys among the users of the system on their own account. However, the research group is encouraging feedback from the information providers during the entire trial and afterwards.

Also working on behalf of the Federal German Post Office (Bundespost) is the 'Working group of applied system analysis' (ASA) in Cologne. In April 1979 this group commissioned the research group Hammerer, in Munich, to conduct a 'Study to ascertain the structure, spectrum and potential of commercial teletext usage'. Through this they hope to gain an overall view of up-to-date planning by information providers and an estimation of development potential for videotex.

Legal questions

Bills for videotex laws have been presented in the Land Parliaments both in Düsseldorf and in Berlin. According to this, a trial in Düsseldorf will only be possible for a maximum period of three years. The Düsseldorf law provides for a 'reliability test' of the information provider. News must be 'objective'. Advertising is possible, as a ban would be unconstitutional. The ruling here has been left vague; possibly all that will be necessary will be a label to denote advertising material.

The question of advertising will be dealt with more strictly by the Berlin law. However, there will be no 'reliability test' of information providers in Berlin, although in cases where there is a gross violation of the regulations the licence to use videotex could be withdrawn. The Post Office would furthermore have the duty of offering an alternative version if the supplier should cease to exist. Furthermore, the bill envisages a videotex authority who would make it possible for otherwise illegitimate or unlawful suppliers (e.g., minors, persons with immunity, etc.) to have access as suppliers. Services provided by a third party upon entering the system are to be paid for according to a fixed list of charges.

Participation of the press

The press in the German Federal Republic, represented by the Bundesverbund Deutscher Zeitungsverleger (BDZV) (Federal Association of German Newspaper Publishers) in Bonn, and the Verbund Deutscher Zeitungsverleger (VDZ) (Association of German Newspaper Publishers) also in Bonn, have from the very beginning of the development of videotex made a strong claim to participate in, and make professional use of the new system of text transmission. The newspapers and publishers of periodicals represent the point of view that the transmission of teletext or videotex does not fall under the definition of broadcasting, but clearly represents press publication. The publishers regard teletext and videotex as an 'electronic supplement' to their conventionally printed publications. Some authors (for instance, this author, in 'Die Bildschirmzeitung'—'The Telenewspaper'—'Read television rather than watch television', *Colloquium Verlag*, Berlin, 1977, p. 79) make the assertion that teletext and videotex fulfil the general concept of a newspaper, assuming that a medium can be classified as a 'newspaper' if it has the following characteristics:

- periodic appearance
- mechanical reproduction
- public appearance, i.e., accessible to all
- variety and comprehensive coverage
- universality of interest
- topicality
- commercial production.

According to this point of view, not only are these characteristics fulfilled by videotex, but it is furthermore possible to print out the texts that have been recalled on the television screen at will by means of printing apparatus, thus materializing the text just as in a newspaper.

All local papers in Düsseldorf/Neuss will participate in the videotex trial—including the business daily paper *Handelsblatt* and as the national papers *Frankfurter Allgemeine Zeitung* and *Welt*. The local *Rheinische Post* will work together with the local *Westdeutsche Zeitung*, the *Neuss-Grevenbroicher Zeitung*, and the *Express-Düsseldorf* to keep down costs. This kind of editorial co-operation is, however, open to all other local or regional newspapers.

The publishers of newspapers and periodicals in the Federal Republic do not fear the new medium as a journalistic publishing rival, but they do see it as a threat that might allow non-newspaper, non-periodical elements into the advertising business. As the videotex system is well suited to the publication and distribution of advertisements arranged in columns and headings, a new form of text transmission for this kind of advertisement could prove a threat to the existence of some papers. This applies above all to the national newspapers who are, to a large extent, dependent on job advertisements which could, again, be displayed well, quickly, and cheaply via videotex.

The market for the sets

The German producers of television sets held back in the field of videotex until the radio and TV Exhibition in August 1979, due to uncertainty in media politics. However, at the 1979 Broadcasting Exhibition hardly a single TV set was on display that had not been prepared for the reception of videotex. The producers, Siemens, Blaupunkt, Philips, and others were offering, by the end of 1979, TV sets with videotex decoders at a price of about 4000 DM (£1000). With mass production the industry reckons that a set equipped with a videotex decoder will cost about 10 per cent more than a set without these additions.

16. France

Hervé Nora

What is videotex?

Videotex is the use of the television screen to display data, information, alphanumeric messages, or graphics. Videotex can be interactive (as in the French Teletel system) or broadcast (as in the Antiope service of TDF, the French television transmission authority). The French proposal on harmonization (the Antiope standard) ensures compatibility between interactive and broadcast videotex thanks to the simple addition of a Didon card to the interactive videotex terminal. This terminal can take two forms:

1. An assembly containing decoder, modem, and alphanumeric keyboard which allows the domestic television set to be connected to the telephone network: this assembly can be integrated into the television set, or not.
2. A special terminal integrating a TV screen, decoder, modem, keyboard, and eventually the telephone; this is, in particular, the terminal for the electronic telephone directory (described below).

There are several justifications for this special terminal:

- the telephone and the television set are generally put in different rooms
- having only one screen to view television programmes as well as, for example, consulting the electronic directory, would pose numerous problems
- given mass production, the cost of a directory terminal with a black-and-white screen about 20 cm in diagonal, should settle at between 300 and 400 francs (plus tax)

Whatever its form, the videotex terminal will constitute a new tool for communication, opening up possibilities of dialogue developing with other terminals and with information services of every kind (and particularly databases).

The electronic directory

In 1983 the 250 000 telephone subscribers in the Ille and Vilaine region of France will be given an 'electronic directory,' in the form of the special videotex terminal described above, which will be installed free in their homes. Using it, they will be able to look up the names of all other telephone subscribers, at first in Ille and Vilaine, and later in the whole of France. Equally, they will be able to consult the 'professional list' (equivalent to Yellow Pages), and have access if they want it (that is to say, if they ask for it) to supplementary information of an advertising nature. This information will be the substitute for the real advertisements in the printed directory, and will be in a form and governed by rules not yet defined.

    The electronic directory should bring considerable improvement in the subscriber's ability to make directory enquiries, notably thanks to weekly and, eventually perhaps, daily update of the directory index. Technical feasibility is established today, but numerous obstacles remain, particularly in the area of 'man–machine' dialogue. It is imperative that this service be able to be used, without special training, by at minimum all those who today know how to use the printed directory. In this respect, the design and ergonomics of the keyboard will without doubt play a very important role.

If the trial in Ille and Vilaine is a success, the service will be progressivly extended to the rest of France over a dozen years; by 1992 about 30 million electronic directories would then be in use. This development will be financed:

- by the economies realized on the printed directory (forecast cost in 1983: 50 francs per subscriber per year)
- by a tax on the local communications cost incurred in consulting the directory.

This tiny surcharge should be easily accepted by the subscriber in view of the clear improvement in the service offered to him. For an average subscriber who consults the printed directory once a week, it would represent an annual charge of 25 francs.

The Velizy trial—project Teletel

The collection of services (other than the electronic telephone directory) that can be accessed by a videotex terminal will be tested from the end of 1980 in the Velizy area, consisting of the towns of Velizy, Versailles, Buc, Jouy-en-Josas, and Les Loges-en-Josas. This trial, which will involve about 2000 to 2500 households, has as its objectives:

- to try out the services, their appeal to the public and their acceptability
- to study on a realistic scale th economic and social consequences of each of them

The trial is made particularly important by the fact that it is impossible to study 'on paper' the consequences of these services for diverse economic activities and for the adaption, or otherwise, of existing regulations governing, in particular, information, competition, and advertising.

The user of a videotex terminal seeking to get himself to Japan (for example) will not call up the term 'Travel' from a central databank. He will call up Air France, or Jet Tours, or more probably one of the local travel agencies. Such agencies, like the smaller information providers in general, will make use of an intermediary to handle their data, who could be for example, a newspaper publisher. To take a concrete example, one can imagine that the travel agencies in Velizy and Versailles will communicate with their normal enquirers in search of information and advertising, by using the *Versailles News,* for example. The subscriber himself will call up the term 'Travel' from the service offered by this newspaper, which could also affer among other things classified advertisements, local news, etc.

In order that all the potential information providers shall be able to participate in the Velizy operation, a computer centre run by the Ministry of

Telecommunications will be put at their disposal for the duration of the trial. Its use will be voluntary and the centre will disappear at the close of the trial. The cost of the trial will be apportioned in the following way:
- the subscriber will find his communication with the Teletel centre taxed on a time basis (probably per 5 minutes of usage) and the service itself will be priced to the user at a level decided by the information provider; this service could eventually be free of charge
- the IP who makes use of the Velizy computer centre will find himself billed for an amout proportional to the machine capacity which he needs
- the IP who does not use the Velizy centre will be accessed through the Transpoc network: the corresponding tariff rates will apply.

At the time of writing, some 200 companies or organizations expect to offer a service in Velizy by the end of 1980. By way of example:
- SNCF, the state railways, will offer a seat-reservation service, its customers being able at their choice to go and claim their tickets at the station, or at a travel agent, to get them through the post
- several banks will offer their clients the ability to check from home their bank account and the last ten entries on it, along with a facility for automatic payment of bills. This service will mean the addition of an identity-card reader (probably a magnetic card)
- a message-switching service will allow subscribers to communicate both between themselves and with their suppliers
- several mail-order houses will offer an automatic ordering service and, in conjunction with the banks mentioned, a system of payment linked to the order
- using as an intermediary the CEESI (the study and Development Centre for Information Systems), which will ensure co-ordination under the direction of the General Secretariat of the Government, the civil service ministries and other public and semi-public organizations will put at the user's disposal essential public information (current regulations, rights and procedures, etc., perhaps to more than 20 000 pages) as well as a message service connecting with the Interministerial Centres of Administrative Information and the principal external services involved
- the French telecommunications ministry will offer an enquiry service on taxation, installation of communications equipment, etc., and a way of paying telephone bills. It will also offer to assist users of the system and maintain a directory of the information providers.

In addition 1000 of the 3000 sets used in the trial will allow access to the Antiope services offered by the TV networks, and distributed by TDF.

Preparation and conduct of this trial will be carried on in close liasion with interested financial partners, the chosen locations, and obviously with the population of the areas involved. A permanent press-telecommunications working party was set up in June 1979 to study problems raised by the development of this new medium.

After Velizy

At the close of the Velizy trial (say, towards the end of 1982) the lessons drawn from it will enable a decision to be made about the services which could be offered to subscribers having a videotex terminal (that is to say, subscribers who have been given the electronic directory *or* those who have bought the directory terminal, a videotex television set, or an adaptor for using their normal TV set). The conditions under which these services will be offered will be settled at the same time. Any necessary legislation and regulations will have to be made. These services will then be offered:

- by companies or organizations having their own computerized information, by direct access (these computers could be nothing more than a personal computer, in the case of a trader, for example)
- through computer service bureaux, acting as intermediaries, in the case of those companies not wishing, or not being able, to put their own equipment into operation.

On present estimates, it seems likely that local services will be called up via the local telephone network; while for obvious reasons to do with cost of transmission, it will doubtless be necessary to use the Transpac network to gain access, for example, from Limoges to the SNCF reservation service in Paris.

Special access points to the Transpac network will perhaps have to be installed and will constitute a Teletel 'network'. It is however possible that future Transpac 'bottom of the range' switching points will be sufficient to fulfil this role. The services will be listed in the electronic directory of the telephone subscribers, on terms not yet defined.

The subscriber will be billed:

- by the telecommunications authority, for the cost of his telephone usage
- by the information provider, for the cost of his service. One can forsee the emergence of specialized billing companies, to the extent that use of electronic payment systems, and in particular magnetic cards, do not produce an elegant solution to this delicate problem.

Business videotex

In addition, as soon as videotex terminals aimed at the general public are entering the market in sufficient quantities—that is to say, from the end of 1980—companies or organizations will be able to set up or extend existing internal internal information networks (telephone networks and data transmission networks) to which can be connected these same terminals, or near-versions of them adapted to specific professional applications.

Numerous applications, whose extent is still today limited by the cost of terminals, could be offered at the same location or between different establishments. Among these applications one could cite:

- scanning files
- management (stocks, orders, deliveries, billing, accounts)
- internal electronic message switching.

There are a lot of organizations already working on these professional (that is, specific, applications of videotex.

17. The Netherlands

Folgert de Jong

Viewdata was introduced to people in the Netherlands at the international 'Firato' Radio and TV Exhibition in Amsterdam in September 1978. The new medium was displayed by the General Post Office (PTT) and the Philips electronics company. On separate stands they showed the 4000 frames which, for display purposes, had been fed into the data bank. On a third stand Dutch newspapers together with the Netherlands news agency ANP combined to demonstrate how these pages are fed into the system and how they are kept up to date.

Before this public demonstration, Philips and the PTT had passed through a phase of research. To bring viewdata to the Netherlands the PTT had bought the system from the British PTT. Philips developed a system of its own. The research work and the demonstration at the Amsterdam exhibition took place on a 50/50 basis. When Firato was over, Philips went its own way and developed a system of its own, some examples of which have already been sold, primarily in South America.

The PTT remained faithful to the system bought in Britain, and the hardware and software of the British GC 4082 was adapted to Dutch demands. Firato's success had been an added reason for the PTT to go on developing viewdata in the Netherlands. The decision was taken to start a nation-wide viewdata trial in the Netherlands during 1980 for a period of one year. Subsequently the trial will gradually turn into an operational service This schedule was fixed in consultation with the Interdepartmental Viewdata Steering Group, set up to advise the government on viewdata and teletext.

The steering group comprises representatives of Dutch publishing houses, the broadcasting system, and the Ministries of Education, Justice, Home Affairs, Culture, Recreation and Social Welfare, and Transport and Waterways. In the Netherlands broadcasting comes under Culture, Recreation and Social Welfare and the PTT under Transport and Waterways. The Home Ministry is represented in the steering group in view of the privacy aspects of viewdata in particular. The use of the two new media in the Netherlands has not yet been regulated by law. However, indications are that teletext will mainly go to the broadcasting system while viewdata is expected to become an open medium with complete freedom of participation. The central data bank will be entrusted to the PTT, which has called its system Viditel. No legislation so far been presented which is opposed to private ownership of viewdata banks.

One private viewdata operator has already come forward. It is Applied Viewdata Systems (Toegepaste Viewdata-Systemen—TVS), a subsidiary of the publishing house Intermediair. TVS will get a computer for use by closed-circuit consumer groups. TVS's viewdata system aims mainly at selected groups. Feasibility studies are centred in the medical community, insurance, the food trade, the non-food trade, the book trade, and the education sector.

The PTT, which also allows for closed-circuit consumer groups in its viewdata system, is not opposed to such an independent TVS Viewdata system, It argues that the TVS system will bring more consumers into the market to the benefit of the entire medium.

Immediately after the green light was given for a trial, a special PTT committee started developing a viewdata search-tree, adapted to Dutch requirements. In addition, a survey was started to get some insight into the information requirements of potential viewdata users. Special attention was paid to the various types of information required and the locations where—and periods at which—the need would arise for each piece of information. The PTT had this survey carried out by an independent market-research agency. One of the surveys was held in trade and industry and a second among households. In all, 753 firms took part while in the private sector 1264 people kept a diary for weeks to determine their information requirements. It stands to reason that this survey should not be regarded as a clear verdict on the need for information through the viewdata medium, but potential information providers (IPs) at least have some insight now into the demand there is for information. The PTT makes the findings of the study available to IPs free of charge.

The PTT will also accompany its viewdata trial with an enquiry on its costs. The findings will be forwarded to all IPs. In principle, these findings will be made public, but if any participant wants his share in the experiment to be treated confidentially, the PTT is willing to do so, on payment of a consideration. Whenever an IP wants additional information the PTT is willing to incorporate relevant questions in the enquiry, but costs will be charged to the IP. The PTT prohibits separate enquiries by the IP himself for fear of 'over-taxing' people taking part in the trial. The IP will, however, be allowed to hold an enquiry among the users which he himself brought into the test.

During the Dutch viewdata trial, each provider of information will have to contribute his 'own' users. Learning from Britain's problems, the PTT has designed a system to ensure that there are sufficient users at the beginning of the test, so as to avoid a vicious circle like the one that slowed down Prestel in Britain. The argument advanced by the PTT is that IPs will be prepared to incur costs only if there is a sufficient number of users. Users, on the other hand, will not buy a viewdata set or subscribe to viewdata until sufficient information is provided. The electronics industry too, is playing a major part. It will not start producing sets in large numbers until sales prognoses are favourable.

To avoid the 'British disease', the PTT intends to break the vicious circle in a joint effort with the IPs. The PTT is offering the IPs its computer for the test free. Pages used and subscription fees will not be charged. For the input of information, however, a computer rate of ten guilder cents per minute will be charged. The PTT requires IPs to contribute one user for every 50 pages, or part thereof, which they use. An IP using 1000 pages, will have to install 20 viewdata sets or decoders with users. IPs will have complete freedom as to the choice of users and financial arrangements with them. But they will have to take part in the PTT market research. The PTT is prepared to assist in installing the sets, if its assistance is required.

IPs have the advantage that they need not provide the modem (the modulator – demodulator that connects the TV set to the telephone system) as the PTT will lease modems to users as part of the viewdata subscription. The subscription rate, including modem rent, will be 10 guilders a month.

During the test the cost to IPs will be as follows:
–   contributing one user for every 50 pages or part thereof
–   possibly lease of editing terminal, f 5000 a year
–   telephone costs (14 guilder cents per $\frac{3}{4}$ minute)
–   use of data bank (10 guilder cents per minute).

The cost to users during the test is:
–   purchase of viewdata set (possibly in co-operation with the information provider), f 3000 to f 3500
–   annual subscription, $12 \times f\,10 = f\,120$, including rent of a modem
–   telephone costs (14 guilder cents per impulse)
–   use of data bank 10 guilder cents per minute
–   information cost per page.

After August 1981, when the Dutch viewdata test ends, the cost to the user will remain the same as it was during the test. The cost to the IPs, however, will be as follows:
–   subscription of f 10 000 per year
– f 10 per frame per year
–   possibly rent of editing terminal, f 5 000 per year
–   telephone costs
–   use of data bank, 10 guilder cents per minute.

The costs to IPs and users of closed user groups will be the same as for public services. There will be no market research on such methods of information provision. Publimedia BV and the Netherlands Realtors' Association were the first IPs to use closed user groups. Publimedia will use viewdata only for internal communication purposes for the time being. The Realtors' Association sees viewdata as a start of a swifter exchange of information between its computer centre and its members throughout the country. Both of these IPs will also take part in the public viewdata test.

The development of closed user groups is helped by the marketing of a small-screen desk terminal with a built-in decoder and modem. The PTT is leasing these sets of subscribers to viewdata for f 120 a month. Participation in a closed user group via the PTT is advantageous because such subscriptions entitle users to use also the public viewdata network. The same subscription rate covers access to

199

closed user groups and the public data bank. Users have access to these various kinds of information by the same cliënt and code numbers. The viewdata test involves a problem in that the PTT is unable to introduce a national rate for the cost of the telephone connection with the viewdata data bank before 1983. The data bank is in The Hague, so that residents in that city and surrounding district will be able to contact the computer for 16 guilder cents (soon to be reduced to 14 cents) irrespective of the duration of the call. In other districts, callers will have to pay the high daytime rate of 14 cents per impulse, lasting approximately one minute.

Several IPs have grave objections to this differentiation, which in their view makes it impossible to carry out effective market research. The PTT has accommodated the providers by exempting subscribers outside the Hague district from the obligation to pay the data bank rate of 10 cents per minute in the daytime (from 8 a.m. to 6 p.m.). The exemption will be reconsidered in May 1981. Outside these hours and during weekends, subscribers will be charged only 5 cents per minute. Subscribers within the Hague district will have to pay 10 cents computer costs.

The PTT is trying to make it attractive also for small IPs (fewer than 70 pages) to build up an information file in viewdata. Every big IP is allowed to 'sell' to smaller providers pages leased from the PTT. In this case big providers will be acting as an 'umbrella organization'. Four such organizations are:
1. Publimedia
2. Baric, a subsidiary of ICL and Barclay's Bank
3. Fintel, a subsidiary of the *Financial Times* and Extel
4. The PTT is also acting as an umbrella organization, charging an annual rental of ƒ 150 per frame instead of an annual subscription rate.

The test uses a double GEC 4082 carrying 120 000 pages which can be expanded easily. The purpose of the double computer is avoidance of stoppages in the event that one computer breaks down. For feeding and obtaining information, 208 telephone lines are to be available. The PTT plans eventually to have Viditel data banks in several parts of the country.

The PTT is reckoning on some 4000 users at the start of the market test. There cannot be more because otherwise they would find the line busy too often when dialling the computer. The big question remains whether industry will produce the required number of sets, as otherwise the Netherlands would still be confronted with the feared 'British disease'. The PTT is fairly confident that the sets or decoders will be available in time. One of the chief future IPs, Vereniging van Nederlandse Dagbladuitgevers (Netherland Newspaper Publishers' Association), is having talks with industry to see to it that the required number of sets will be available. The daily newspapers have their own plan for joint participation in viewdata. They expect to require 5000 pages, with a claim on another 5000, which would mean a contribution of 100 user sets to the system.

Under the joint newspapers' plan which uses the name Krantel, the national news agency ANP will provide a general news service for viewdata. Every newspaper may take part in the test independently and under its own name, under the umbrella of the association. Newspapers wishing to await the results of the test will be able to let journalists gain practice at a central organization. The newspapers will jointly finance the general contribution towards the system. These costs will be shared on the basis of their number of subscribers. It is intended to insert adflashes on the editorial pages which will refer to the advertisement pages. To this end a special division will be set up for the service. The newspaper itself will insert the adflashes on its own 'viewdata newspaper'. These flashes and advertisements on viewdata may again be linked to advertisements in the newpaper.

In the Netherlands, too, the start of viewdata is still more or less a leap in the dark. How much so will not become clear until August 1981, when the test will have to change smoothly into an operational service.

18. Finland

Jaakko Hannuksela

Finland has a mixture of private telephone companies and a PTT operating in the telephone and telecommunications field, the emphasis being on the private side. There are about 60 private telephone companies in the country giving telephone service to about two-thirds of the subscribers. The PTT takes care of long-distance traffic and international calls. At the same time it co-ordinates the intertwined private systems with technical standards.

Radio and TV service in the country is given by a government-owned body but the press is private and very diversified. Helsinki alone has seven dailies with national distribution. The press has fulfilled expectations rather well in the past. A proof of this is the fact that a parliamentary committee back in 1972 stated that 'the electronic distribution of newspapers with terminals or/and facsimile is not broadcasting', and should be reserved for newspapers.

In this relatively liberal environment, Sanoma Publishing Company saw back in 1976 an opportunity to expand its field of activities. It invited Nokia Electronics and Helsinki Telephone Company to join forces in a viewdata system. After preliminary studies a public trial was started on 19 June 1978. After evaluating 'make' or 'buy' alternatives, a 'make' option was chosen. This was mainly due to hardware and software maintenance reasons in a relatively remote country. Prestel standard was chosen because it was judged to be close to the coming European standard in videotex. The organization of the project is small, only five persons, but it is backed up by the resources of the sponsoring bodies. The total expense till the end of March 1980 was about 500 000 dollars.

Who is who?

The Sanoma Publishing Company is the largest publisher in Finland with a turnover of 130 million dollars. Its newspapers are *Helsingin Sanomat* and *Ilta-Sanomat* with a total circulation of more than 500 000. Aside from newspapers, Sanoma has several magazines and printing plants in the Helsinki area.

Nokia Electronics is a division of the diversified Nokia Group, which has activities in the fields of cable, rubber, pulp and paper, etc. The electronics division has long experience with terminal, computer, and telecommunications-equipment manufacture.

Helsinki Telephone Company is a co-operative owned by telephone subscribers in the greater Helsinki area. It has been granted its operating licence by the Government Telecommunications Department, not the PTT. In actual fact it operates like a private company, forming its own operating policies.

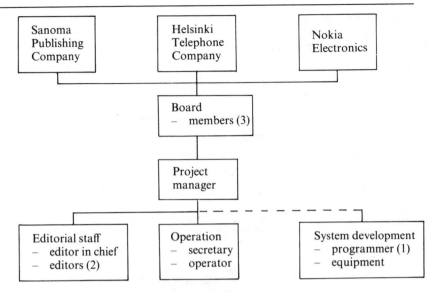

Fig. 18.1 Organization chart

The general principle of the project has been that the costs and risks should be equally shared by the partners. The main responsibilities have been divided according to the normal fields of operation of each partner. Sanoma Publishing Company is responsible during the trial for the supply of information in the system. The Helsinki Telephone Company has supplied the telecommunication equipment, data exchange, modems, and connections to the telephone network. Nokia Electronics has supplied the computer and business terminals which can also be used as editorial terminals. A daughter company of Nokia, Softplan, has been responsible for the software side of the project.

Technical design

In short, the system is functionally identical to Prestel as seen from the subscribers' point of view. Because Nokia has long experience with POP 11 computers, a POP 11/34 was chosen for the project. The operating system is RSX 11M, which is optimized for fast response to real-time events. At the same time it enables software development to be carried out simultaneously with normal operation. The software is upwards-compatible with other computers in the same series so that enlargement of the system can be done smoothly in the future.

Home terminals are colour TV sets manufactured locally by Salora with built-in decoders. Business terminals are black-and-white and manufactured by Nokia Electronics. For the time being, standard 1200/75 bit/s modems are used before a compact, dedicated and cheap modem is ready.

The present software is functionally identical to that of Prestel except for networking facilities between computers. In the future either Dec-Net or a dedicated packet network will be used to tie local computers together.

Style of database

The present database uses about 10 000 frames of the total capacity of 50 000 frames available. The main headings follow the Prestel example closely (see Table 18.1).

Table 18.1 Database organization

1. News and weather
   News
      Local
      Domestic
   Sports
      Sports news
   TV
   Helsinki cultural events
   Games, pools, etc.

2. Home database
   Home and family (incl. meal trips)
   Education
   Transport

3. Business Teleset
   Business and finance news/weekly
   Finance markets
   Economic development
   Government
   Company report database
   Who's who

4. Telset guide

5. Alphabetic index to pages

The main database is still organized in a tree-like manner. During the past few months more and more cross-referencing from a page to a related page in a different branch has been used. In this way the structure has started to grow in a complicated manner. This means that there is, for example, direct optional access from local news pages to events calendar pages etc. The role of Sanoma Publishing Company has been dominant as information provider. There have been several active information providers in the project:
- Telset-newspaper staff
- Sanoma Information Service
- Government Statistical Office
- Government Technical Research Centre
- The City of Helsinki
- Gallup
- Largest banks

The future

A company, Teletieto Oy, was established in January 1980 to give videotex network service in the Helsinki area. It will operate on a common-carrier basis. Sanoma Publishing Company will publish an electronic newspaper in the database but every interested information provider can act as an information source on equal terms. The information providers will be required to observe publishing laws.

The intention is to apply in the public-service pages mutual principles stipulated by an agreement between the network company and the information providers. All index pages are free and the user is informed in advance of the price of the page. Information providers may freely set the prices on their database. In the first three years of operation the system will be especially of interest to companies and collective bodies. The information providers will therefore in the beginning concentrate on composing a database for business use.

The market projections are as follows:

Year		Terminals
1980	200	Business terminals
1984	5 000	Mostly business terminals
1990	100 000	Mostly home terminals

All Finnish newspaper publishers and private telephone companies have been offered the technology practically free of charge. We try to promote the idea of local newspapers and telephone companies establishing their joint Telset ventures. Standards will be identical and actively enforced.

19. Japan

Hisao Komatsubara

The Captain System (Character and Pattern Telephone Access Information Network System) as a test trial started on 25 December 1979. This prompted Koichi Yasuda, Director of the Telecommunications Policy Section, at the Ministry of Posts and Telecommunications, to say 'It is an information present to the 1980s.' The experiment will be carried on at least to the end of fiscal 1980 (ending 31 March 1981), and is organized jointly by the Ministry and the Japan Telegraph and Telephone Public Corporation (NTT) with a total of 1000 receiver terminals selectively installed in households in Tokyo, but including those lent to the information providers. At the beginning, there were 165 information providers participating in the system and they are expected to input a total of 100 000 pages of information into the Captain computer, when the trial becomes full-scale.

Unlike Great Britain, where the British Post Office runs Prestel by itself, a special legal entity by the name of the Captain System Development Institute (Captain Centre) was created in February 1979 to administer the experiment. The reason for this is very simple: the government as a whole is strictly forbidden to increase personnel beyond the number stipulated by law, and the Postal Ministry usually lacks the funds to operate such an experiment anyway. Instead it is the NTT which supplies both money and expertise to the Captain System. The NTT is a huge public corporation with an annual gross turnover of 4000 billion Yen, with its profitable telephone service and money-losing (but minor in scale) telegraph operation, whereas the Postal Ministry has to look after an enormously unprofitable postal service. The NTT for the current fiscal year is expected to yield a gross profit of 400 billion Yen, and is an ambitious monopoly always searching for ways of expanding the scope of its franchise, from data telecommunication to satellite communication. It is both a common-carrier and a centre of technical know-how exercising hegemony over a number of telecommunications equipment manufacturers. The establishment of the Captain System Development Institute was in a way an attempt to neutralize the power of NTT, though it has been mainly responsible for Captain's development and funding, (1.8 billion Yen has been invested so far). Nobody is yet certain what will happen to the Japanese videotex system when the Captain test service is over and it goes into commercial service. Will it be the NTT who operates the commercial Captain System, or somebody else such as a joint stock company composed of information providers? This important question remains to be solved while the test trial continues, along with other questions such as overall tariffs for the information providers and for the users of the system.

When the trial service began, it did not start with the logistical precision commonly observed in Japanese ventures. On the contrary, it was delayed by at least five months after the target date originally announced by the Ministry two years ago, when it disclosed for the first time to prospective IPs its intention of staging the Captain experiment. In addition, there were little more than 300 terminal units in operation and just about 15 000 pages of information stored on the central computer, although subsequently terminals were installed and pages increased to reach the set target. The delay in terminal installation was not due to the failure of manufacturers to supply enough units, but to difficulty in arriving at agreement between the management and the union at the NTT over conditions of work, since Captain was an addition to the normal workload of the union's members. As far as the supply of terminal units was concerned, there was no problem, since there is a co-operative relationship between the NTT and its equipment suppliers.

On the part of the Postal Ministry, there seemed to be an imperative urge to start the Captain trial before the decade of the 1970s closed. Obviously, the British Post Office's success with Prestel, success in starting if not in operating it properly, prompted telecommunications officials at the ministry to ensure that the trial began when it did. It is no secret that they want to perfect Captain so as to be adaptable to those countries speaking ideographic languages, *vis-à-vis* viewdata that is more suitable to alphabetic languages.

Captain is specially designed to display patterns and Kanji ideographic characters (Chinese characters adopted in Japanese). Kanji characters are not only complicated in shape but large in number even in daily use, ranging from 2000 to 3000 in number. This in itself tends to impede technical application of sophisticated electronics and computerized technology. For European and other alphabetic languages, in which sentences can be written with a comparatively small number of characters and codes, viewdata may be suitable with its code transmission system. By means of the character generator built in to the terminal, coded information transmitted from the database is easily and economically converted into characters and patterns. This transmission system is unsuitable for the Japanese language, as it would require a very large character generator installed in each terminal. For this reason, a quite different method has been developed for the Captain System, the pattern transmission system. Instead of being in the terminal, a large-scale pattern generator is installed at the centre for shared use, and the signals describing the patterns of characters and figures are directly transmitted to the users.

This transmission system adopted by Captain has both advantages and disadvantages. Because images are transmitted in patterns, the system can reproduce much finer-grained displays; there is less corruption of the transmitted characters; and the system permits simplified and less expensive terminal equipment. But it has a major drawback in that the time required for transmission is necessarily longer compared with coded transmission.

The Captain System can display as many as 3500 characters, including Kanji, Katakana (the square form of Japanese syllabary), and Hirakana (the cursive form of Japanese syllabary), in addition to alphanumerics and special symbols. Character display is available in three different sizes; up to 120 standard characters (16×18 dots) can be displayed on one page, but the number can be increased up to 480 characters by using a smaller size (8×11 dots). The use of smaller characters may, however, decrease legibility. Captain claims a unique feature in being able to offer various kinds of display modes, notable among them a continuous scrolling of characters either upward or downward, extending to more than one page. The user does not need to push a button repeatedly for continuous information, and he may freeze the display at any time he wishes by operating a stop button on the keypad.

It may seem strange to outside observers, but the trial service of Captain is being conducted free of charge to any of its participants. The user is not required to pay anything for the information he receives page by page, except for use of the telephone by which he retrieves the stored information from the Captain computer; nor is the IP required to pay entry fees or charges for the use of computer storage (but neither does he get any income from the users). Thus, neither the NTT nor the Captain System makes any financial gain during the test trial. In other words, all three parties are contributing in their respective ways: the public by incurring telephone bills, the IP by covering extra payrolls, and the Captain System by budgeting for initial investment and experiment costs. As noted earlier, the NTT has so far expended only 1.8 billion Yen, for there has been little equipment expenditure except for terminal units. Captain's computers have been housed at the NTT and are in shared use with another NTT project called Demos.

In fact, Captain is but one of many new services planned by the NTT for what they call 'the post-telephone era'. There are over 50 million telephones, including 36 million subscriber telephones, and 7 out of every 10 households are now equipped with a telephone, which shows telephony has reached saturation point. After telephony, what will come next? Captain is one thing and the Video Response System (VRS) is another. The VRS, also under experiment with 100 terminals in Tokyo, is a more ambitious project linking the television set with the central computer via broad-band cable for the transmission of a much greater variety of information in voice, still pictures, and moving images on request by the end user. For this and other experimental projects, the NTT is prepared to expend initial investments to ensure their take-off.

Thus the Captain experiment is being conducted free from tariff problems for the partners in the system. It will perhaps give clearer results as to what sorts of information are preferred by the public viewers, and the IPs in the meantime are expected to gain expertise in editing and display techniques, while the Captain Centre engages itself in system improvements. The IPs are not yet at all sure just what kind of information would best fit this new medium, or what the format and expression for it should be. What they know is that Captain is neither television nor newspaper.

Among the 165 IPs, there are 23 newspapers and news agencies, 30 magazine and book publishing houses, 8 broadcasters (including NHK, Japan

208

Broadcasting Corporation), and 22 advertising agencies. They are the participants from the traditional communications media. But, in addition, there are department stores, travel agencies, public transport companies (including Japan National Railway and Japan Air Lines), banks, and some government agencies and public corporations, all of whom have traditionally depended on existing media. Thus the list demonstrates a mixture of media professionals and laymen, laymen in previous media but equal and competing partners in the Captain System. This fact alone presents two immediate questions, should this new medium become fully fledged. First, these 'laymen' have been in varying degrees advertising clients for the traditional media. Will it mean decreased advertising expenditure for newspapers, broadcasters, and magazines? Second, these 'laymen' will on the whole utilize Captain for publicity purposes, if not outright advertising presentations. What are the demarcation lines between news articles (as in the newspaper) and advertisements, or between programmes and commercials (as in broadcasting)?

At the moment, the IPs are a loose group, apart from the fact they are all on the same system and belong to an organization of IPs. One-third of them have not input even a single information page as yet, but are waiting to see what sort of system they have joined. The most active in providing information to the system are those from the existing media, led by the newspapers. Typically, there are three to four journalists who have been specially assigned to write and edit information items for Captain. Apart from news of daily topics, however, they frankly admit that they are groping their way in the dark. Besides, they face a problem of slow inputting. Captain has yet to develop an input device which an IP can use in his office for direct input into the system computer. Instead, he has to depend on facsimile transmission or on messengers on motorcylce to send hand-written manuscripts to the centre. At the Captain Centre, there are girl operators working on nine inputting machines from 10 a.m. till 10 p.m. in different shifts. But still they can only input 900 pages daily at maximum, and their work tends to be slow because manuscripts are hardly legible (most likely the journalists' fault) or are not written in the proper format according to instructions. These are rather minor if daily problems, but other problems are not so minor.

One of the IPs from the magazine field is *Pia* a very successful bimonthly guide to art, music, film, and other entertainments, well read by youngsters in town. In Captain, this pop journal presents more or less similar guide information but, one day, one of the girl punchers at the Captain Centre is said to have declined to input what was provided by *Pia*. The items were brief reviews of pornographic films and the words used apparently repelled her. It was a story of rare chastity, but between the Captain Centre and this particular IP, it was tantamount to prior censorship. When Captain was in preparation, it was broadly agreed between the Postal Ministry and the IPs that each IP was free to express anything and that he alone should be responsible for the content of information, subject only to the laws that are commonly applied to print media. On the other hand, there seems to be an implicit consensus among all parties that there ought to be an ethical code, which at the time of writing does not exist. Some media observers had feared that Captain would become an outlet for pornography direct into the home. The case with *Pia,* however, was not an issue of pornography *per se* but a question of propriety in the use of Japanese. Given their lack of experience, Captain officials should not be censured for their conduct. 'What would they do', Hayato Matsui, a *Pia* editor wondered, 'if one of the encyclopedia pages in the Captain database carried a diagram of female sexual organs?'

What code of ethics should be applicable to Captain largely depends on how one looks at this new medium. It is neither broadcasting nor newspaper but an information-retrieval medium to which the user has freedom of access, upon payment. It is not a one-way flow of communication with the freedom of choice entirely in the hands of communicators, as in the case of TV or the newspaper. What kind of standards should there be for such an interactive medium as Captain? It is one of the many problems that will have to be resolved in the study groups that were organized among the IPs.

Earlier, there were misgivings that videotex would be a threat to the convential media, especially the newspapers. A report from England said that one-third of classified advertising would be sucked away from local newspapers when Prestel went into commercial service. But the people in the Japanese press who are now directly involved with Captain are wondering if it will ever become a profitable venture. It may or it may not; no one can tell at this moment. As a medium it is based on already heavily capitalized networks of telephone and television, so no doubt there is little need for that initial investment in equipment and facilities which normally accompanies the start of a new business. On the other hand, videotex as we see it now is a cumbersome process in that the user often finds himself turning several pages of indexes before reaching his needed page. After that process, is the public prepared to pay for information on top of paying the telephone bills? One also has to realize that we are in a society oversaturated with information—the Japanese fondly call it the information society. According to the census of information consumption by the Postal Ministry, the volume of information supply tripled in the years between 1960 and 1975, while the rate of consumption decreased from 40.8 per cent to a mere 9.9 per cent in the same span of years. The conclusion must be that, in spite of an ever-increasing supply of information, a great deal of it has passed unnoticed. From the media point of view, it means that they are increasingly in a competitive market for the public's attention. Videotex, the brainchild of PTTs the world over, is entering as yet another competitor in this crowded scene.

20. The USA

Michael Nyhan, with Robert Johansen and Robert Plummer

The USA enters the 1980s as a novice in the world of videotex and teletext. Normally in the forefront of technological change, US companies have stood back from this new set of technologies, adopting a 'let's wait and see' attitude. In the mid- to late 1970s, while European and Canadian government agencies were releasing press announcements proclaiming their advances in teletext and videotex, the US was curiously silent. Not until 1979 did the topic of videotex and teletext gain any degree of national exposure—and even then it was more talk than action or systematic exploration.

Why this unusual stance? Does the USA know something the others don't? What has the USA been doing in videotext and teletext? How are these systems likely to fare in the USA during the 1980s? The question of why the US has been so slow to join the group of countries pursuing the videotex/teletext phenomenon has both simple and complex interpretations. A simple answer might go like this: in contrast to many European countries, the USA does not have a single government agency, like a PTT, responsible for telecommunications and postal service. The USA lacks a central authority with the ability to orchestrate the many forces and actors necessary to rapidly inaugurate a service as multi-faceted as videotex. Traditionally, the USA relies on the private sector to introduce new technologies. In some instances, of course, the federal government intervenes through massive investment to spur a technology forward (a classic example is satellite technology). For a variety of reasons, the US government has paid scant attention to the development and potential of teletext and videotex.

A more complex interpretation of US sluggishness in this area would take into account a number of factors and changing national conditions. Among these would be:

- increasing emphasis on government deregulation of industry (airlines, trucking, banking, and telecommunications)
- a lack of approval of technical standards for any system (no company wants to be stuck with heavy investment in a system later rejected)
- a fluid policy and regulatory environment (the USA has been re-examining its existing communications policy to determine if it is still viable in the computer-communications era)
- an unformulated and untested market for videotex and teletext (what do people want and what are they willing to pay for?)

Note: The authors are on the research staff on the Institute for the Future in Menlo Park, California.

- a sceptical attitude towards new technologies that purport to bring dramatic changes to the home and business (the same promises were made for cable television in the early 1970s)
- an opinion among some that videotex is unimpressive and simply not at the cutting edge of new information technologies.

Such factors—and certainly one could cite more—combine to delay rather than encourage investment or even experimentation. Whether it is informed hesitation is quite another question.

Who is doing what in the USA?

The manner in which Britain developed and launched its broadcast teletext services, known as Ceefax (non-commercial service) and Oracle (commercial), provides an interesting contrast to the US experience. While both the USA and the UK have shown an interest in providing a television captioning service for the deaf and the hearing-impaired, the two countries are following different paths. In the early 1970s, the BBC began its teletext work with the intent of inventing a captioning service. While the service for the deaf initially proved difficult to implement, the general teletext service was eventually made public. The rationale was simply that a broadcast text service was of benefit to both the deaf and the general community. A more specialized captioning service could be added later.

The US position has now become almost the reverse. The Federal Communications Commission (FCC) has authorized a technologically limited service of closed captioning for the hearing impaired, funded by the Department of Health, Education and Welfare. The service will be inaugurated in 1980—before any national teletext service is approved and operational. Two commercial TV networks, ABC and NBC, and the non-commercial service, PBS, have agreed to participate. Each evening a certain number of programmes will be broadcast with captions, available only to those with appropriate decoders. Involved in its air tests of teletext, CBS declined to participate in the service, arguing that the government-funded closed-captioning system 'may soon be rendered obsolete'. Those who advocate the closed-captioning system admit its limitations (for example, to a single language), but they see it as a long-promised breakthrough for the hearing-impaired community. In response to criticism that captioning would be quickly outdated by teletext, FCC Chairman Charles Ferris quipped: 'The best is often the worst enemy of the good.' In this case, however, the good may indeed be an investment in obsolescence.

Leading pioneers of broadcast teletext in the USA include a Salt Lake City TV station, KSL, the CBS television network, and KCET, a Los Angeles public TV station. While other broadcasters have undoubtedly been following this area, these three were the only ones who asked for and received FCC approval for experimental broadcast as of late 1979. The Electronic Industries Association subcommittee on teletext has been working on developing standards that eventually could be submitted to the FCC, thereby opening a rule-making procedure.

Teletext represents a strange frontier for many broadcasters. Even with intense

212

competition in major US markets, pushing TV stations to have the latest piece of electronic news-gathering gear and the most exciting line-up of programmes, remarkably little attention has been given to teletext. But broadcasters' ambivalence towards the new service is understandable. If a commercial broadcaster cannot guarantee a set number of viewers for an advertiser at a certain time, the station loses revenue. A teletext service could potentially draw audiences away from commercial messages, thereby losing revenue for the station. At the same time, it is not clear how much if any revenue such a service could attract. Non-commercial broadcasters do not face the problem in the same way—with the possible exception of losing viewers during their appeals for funding.

The earliest federally funded effort in the videotex area is the system known as Green Thumb, funded by the US Department of Agriculture and the National Weather Service. This experimental system aims to provide information directly useful to farmers (e.g., weather, crop, market information). The farmer can attach the Green Thumb box to his television set, telephone a central computer, and request certain information, which is sent in a burst over the telephone line for storage in the local memory of the box. The farmer can then hang up the telephone and page through the information at his or her own leisure. Green Thumb, however, is still a long way from general usage. Two hundred homes in two Kentucky counties are to test the system, but the project has had start-up problems and remains uncertain. As with the closed-captioning project, there are basic questions about the way the trial has been organized. The Green Thumb test might be remembered more for what could have been done better than for what was actually accomplished.

On the commercial side, the Telecomputing Corporation of America caught the eyes of many videotex watchers with its system called The Source. Billed as an 'information utility', The Source provides features and services talked about for videotex systems in the future:

- low-cost access to information ($2.75 per hour for 6 p.m.–8 a.m. usage over a nation-wide time-sharing system)
- information retrieval (users access the UPI newswire, travel, weather, educational information, etc.)
- entertainment (games, horoscopes, etc.)
- transactions (The Source offers a discount service for shopping)
- messaging (electronic mail).

Thus, in the late 1970s, while the giants of the industry and the top-level policy-makers were debating the obstacles to introducing videotex in the USA, a small unknown firm found a way to do it. By using home computers, The Source does not now rely on television with its associated regulatory hurdles. But using computer terminals means that the first users are computer hobbyists rather than the general public. Video terminals are currently available from The Source for as little as $625. The Source too, however, has experienced organizational problems and is by no means certain to be a valid US test of a full-scale home information system.

Elsewhere in the videotex area, several major actors lined up with tests either planned or underway. General Telephone and Electronics (GTE), the second largest telephone company in the USA, acquired the US rights to the Prestel system in 1979. Also in that year, GTE acquired Telenet, a major packet-switched network. Since GTE also manufactures television sets (Sylvania), it seemed to be in an excellent position to take the lead in the US market. As with many other companies monitoring this field, however, GTE has maintained a low profile.

From the first utterances of the words videotex and viewdata in the USA, the question has been: What will A.T.&T. do? Together with its operating companies, A.T.&T. (American Telephone and Telegraph, by far the largest US telephone company) is clearly a dominant and critical actor in the videotex area. Indeed, its possible role in computer communications was the subject of intense debate during the 1970s. Through what is known as the Consent Decree of 1956, A.T.&T. has been prohibited from providing data-processing services. In recent years, separating data processing from communications has become increasingly difficult; many argue that the Consent Decree should be abandoned. How this issue is resolved will be a key decision point in the 1980s. Meanwhile, A.T.&T. is both keeping tight-lipped and making certain all its options are covered. In 1979, A.T.&T. quietly announced that a field test of an Electronic Information Service (EIS) was being conducted in Albany, New York. Though other services are also reportedly being offered, this service is a test of an electronic telephone directory accessed by way of a small black-and-white TV set.

In addition to the telephone companies and The Source, the other major US industry interest in videotex seems to be coming from newspapers, particularly newspaper chains. The Knight-Ridder company and Harte-Hanks, two major group owners, are actively exploring videotex. Harte-Hanks was one of the early information providers to the Prestel system. Knight-Ridder has gone so far as to establish a separate company called Viewdata of America, much to the chagrin of the British Post Office since there is no relation to the British activities. The new company is planning and implementing a field test involving about 150 homes in Coral Gables, Florida. Reportedly the users will be able to access news, weather, sports; lists of adult education courses; movie, restaurant and theatre schedules; boat and fishing information; and library-type information. Also, they will be able to make purchases by credit card.

When asked why Knight-Ridder was getting so heavily involved in the videotex area, one company executive responded with two words: 'fear' and 'hope'. The fear in the newspaper industry is not so much that they will be put out of business, but rather that electronic newspapers will emerge and provide the type of specialized services that US periodicals have become so successful at delivering to discrete audiences. While newspapers would not disappear, they would clearly be affected. If a successful entrepreneur did nothing but provide an electronic classified advertisement service in an area where videotex had a high penetration, it is likely that the daily newspaper in that town would feel the financial effects. The 'hope' this executive refers to is that the newspaper organization can move from considering itself a paper-based medium to considering itself an information provider over all kinds of media.

In the early US trials, the distinction between videotex and teletext is reasonably clear. However, there are signs that these two classes of technologies may become further hybridized. Originally a teletext test, the KLS experiment moved into a second phase that incorporated a touch-tone telephone call-in capability. Similar mixtures have been discussed for one-way cable television tests incorporating a telephone access link.

Another way to examine the US experience with videotex and teletext is in terms of what has *not* yet been done:

- By the beginning of the 1980s, no cable television company had a cabletext (teletext) channel operating, though many observers think cable is ripe for such a test
- No full-channel teletext, either on cable or broadcast, has been tested.
- No full-scale research efforts have explored the social and other implications of teletext and videotex in the USA
- No serious explorations of the public-service potential of teletext or videotex have begun, though at least two are now under way (the Green Thumb project and a proposal by Alternate Media Center at New York University for a teletext field test with the public television station in Washington, DC)
- No formal government proceeding has been convened to consider any of the regulatory/policy issues
- No systematic evaluation of current tests has taken place, nor has there been any systematic comparison of the various efforts.

Thus interest in videotex and teletext increased dramatically but testing of prototype systems remained sporadic and unco-ordinated. While there were increasing rumours of field tests to be launched, the prevalent attitude in both the public and the private sector was one of hesitancy. Many felt there were still too many unanswered questions.

Some visionaries see the 1980s as the period in which communication advances and information technologies will explode in all kinds of new services to home and business. The sceptics, however, are not convinced; they see many problems and barriers to the revolutionary changes forecast. In the long run, both camps may be right. Before looking at the prospects, therefore, it would be helpful to review some of the barriers that confront US industry and policy-makers.

Current problems

Because videotex and teletext are hybrid systems, they tend to exacerbate policy and regulatory problems that have been lurking under the surface. For example, the FCC has made two attempts to define the boundaries between communications and data processing and to formulate policy accordingly. While the most recent attempt (the Computer Inquiry II) is a step forward, many knowledgeable observers believe that the problem is so complicated that the only real solution can come from Congress. Otherwise, they argue, the entire field will be subject to years of court litigation.

The USA enters the 1980s without a coherent information policy. While Congress and the White House grapple with the development of policy, some thrusts already seem enshrined. One of these is the clear movement for de-regulation. The privileged position of carriers like A.T.&T. has ended. By the mid- to late 1980s, the competition for all kinds of telecommunication services could be intense. In 1979 the FCC de-regulated satellite earth stations and cleared the way for more than one over-the-air subscription broadcast station in each market. Such competition will be especially keen in the business area. New networks, such as proposed by Xerox (X–TEN) and the Satellite Business Systems (SBS), will be providing services similar or equivalent to videotex. Likewise, competition for the access point to the residential market could become increasingly tough towards the end of the decade. The local telephone companies may be competing with a two-way cable service, a satellite-to-home service (proposed by COMSAT), a semi-interactive broadcast service, and perhaps other new carriers.

This new competition will emerge with the blessing of policies favouring less government intervention and more reliance on the market place. As one high-level Carter administration official put it, with no originality but some insight: 'Let a thousand flowers bloom.' Not only does it appear that there will be increased competition for regulated carriers, but a flurry of competitive technologies will inevitably have an impact on videotex and teletext services. Indeed, trying to position videotex and teletext in the market place becomes especially difficult when one takes account of such rapidly changing technologies as: the video disc, which can store thousands of pages of information; satellite-to-home service; personal computers; electronic home surveillance and security; fibre optics; electronic video games; electronic mail; and electronic funds transfer. All of these technologies are currently or soon to be available in some form. Videotex, and to a lesser extent teletext, seems far more uncertain.

If videotex and teletext are to become mass-market media in the USA, the first half of the 1980s may well be the critical period in which companies attempt field tests, propose system standards, and try to put together new services and new markets. One likely result of this shakedown is that many companies will suffer financially. Until standards are set—in both hardware and software— entering this field will be a high-risk business. Venture capital may be hard to secure unless and until a convincing market is established. The battle for standards will be an important one; it could well determine the shape and long-term potential of these media in the USA. As with the colour TV standards selection procedure several decades ago, US consumers must live with what is decided.

One of the more thorny issues facing US policy-makers in the 1980s surrounds the First Amendment, particularly the problem of access. It is not at all clear on which model public policy towards videotex or teletext should be based. Should these new media be modelled after the newspaper, with the US publisher having complete freedom to publish (within the bounds of libel laws)? Should the model be based on common-carrier principles, allowing anyone to use the service who can pay the rate approved by some government agency? Or should the model be broadcasting, providing for standardized government review of a licensee's service to the community? Or to make it even more complicated, should the model incorporate aspects of all of these types?

Difficulties appear quickly. For example, if the common-carrier model were chosen for videotex, all information providers and all information receivers in a particular service area who could pay the charge would be allowed entry without discrimination. What happens, however, if that videotex service is provided over cable TV, a medium not previously regarded as a common-carrier? What about the situation of a broadcast TV station with a teletext service? Does the station owner have complete control over the pages of information that station broadcasts? Does that owner have an obligation to provide any public-service pages on a first-come, first-served basis? These access problems are not easily resolved and will probably be on the public agenda in the USA for years to come.

During the 1970s, particularly after the revelations of federal-government intrusion into the private lives of many citizens, all levels of government made moves to safeguard the privacy of individual citizens. It is safe to assume a pervasive sensitivity to any new technology that might transgress the privacy of an individual. As the USA approaches 1984, new home-information systems will come under close scrutiny.

The success of videotex presumes some sort of 'information market place' beyond anything that has yet happened in the USA. It is not at all certain that information retrieval as a service will be the method by which videotex gains its momentum in the USA. There are many who argue for other evolutionary paths: information retrieval will be one of the later services to gain popularity in the home market; first will come services such as entertainment, security, financial transactions, and messaging.

None the less, whenever and in whatever form videotex is introduced on a large scale, there will be major problems surrounding the billing procedure. The phone systems have sophisticated billing systems, which could be adapted for page access charges. However, if the phone company is not an information provider, it might choose not to allow others to access its billing system, just as it fought strenuously to keep cable television companies from utilizing its telephone poles for attaching coaxial cable. Another problem related to billing points out a sharp difference between European countries and the USA: in Europe the PTTs charge customers a basic fee plus a charge for every call. In the USA 85 per cent of the country has flat-rate pricing (i.e., there is one charge for unlimited calls or a set number of calls within a certain geographic area). Thus in Europe, the PTTs have an incentive to develop increased phone usage; in the USA, there is no such incentive.

The sceptics in the USA typically point to a number of problem areas in addition to public policy. On the computer software side, some argue that the costs of providing the kinds of information of interest to a home market in an easy-to-access, simple-to-use format are enormous. Accessing large databases through combersome and often time-consuming tree structure procedures will be an expensive service only the rich will want or need. In fact, sceptics believe that all the advantages of print media in the USA—the ability to browse and read at one's leisure and convenience—will outweigh the attractiveness of home information systems for most comsumers. Such systems in the home, they argue, are at least two generations away.

A more fundamental reservation involves the entire notion of using the TV set to provide information. People are not used to thinking of television as more than a source of entertainment. Putting text on a TV screen runs counter to what the medium does best, and most US TV sets are already in heavy use for conventional television viewing. Similar questions can be posed about whether videotex will be an acceptable use of the telephone, especially if the phone is tied up for hours at a time.

Current prospects

What then are the possibilities for videotex and teletext in the 1980s? Most close followers of the field in the USA would agree that because of the large number of uncertainties surrounding this area, no definitive scenario or even group of scenarios is possible. Fundamental questions are still unanswered regarding practically every dimension of the phenomenon. Nevertheless, we have some clues as to how these media might emerge:

- *Business vs. residential.* Where are these services likely to catch on first? While the USA has led the world in new developments of computer technology and has provided large corporations with the most advanced systems, significant segments of the small business community have yet to realize the advantages of information systems. This market may be the opening wedge for videotex in the USA. On-line information services to the home on a mass-market basis could be more than a decade away.

- *Information artists.* The entertainment industry is well established in the USA; the information industry is in its infancy. The two industries may begin to merge over the next decade as one attempts to make information attractive and the other to make entertainment more informative. A new breed of 'information artists' seems inevitable.

- *Information packagers.* Organizations that can package information for selected markets could well become the growth industry of the 1980s. Those already in the information business, such as newspapers and publishers, could have a head start in this area—especially if they can separate their corporate identities from a particular medium (i.e., paper). However, new venturesome organizations, such as The Source, may be the groundbreakers in this area.

- *Videotex, teletext, cabletext, or what?* Which of these new media stands the best chance for the immediate future? Some version of one-way cabletext would appear to have a strong possibility. By 1985, cable television should reach 30 per cent of US TV households, according to some analysts. It is a growing medium, anxious to find and develop new services. Since the introduction of pay-TV, cable has rebounded from the financial doldrums of the mid-1970s. Given its excess channel capacity, cable could provide full-channel cabletext service. Furthermore, cable has few regulatory encumbrances that might delay development of a new service. Broadcast teletext is not likely to gain approval at the FCC much before 1983. It could be 1985 before a national teletext service is generally available.

- *Competition from other media and other services.* During its shakedown period, videotex will likely be buffeted about by other new media, such as the video disc, and other new services, such as remote home security. Unless a videotex entrepreneur can flexibly adapt to a changing market environment by switching to different services as needed, videotex may not survive the onslaught of new competition and new services. Or it may continue on a very low penetration scale, as cable did, waiting for some time in the future when the demand picks up.
- *Consumer discretionary spending.* How much will consumers be willing to pay for new services to the home? Business users, it is generally agreed, will pay for services that save time and energy. Household spending, however, is a big unknown. Yet the remarkable increase in the number of households spending money for pay-TV in the late 1970s suggests that the previous stereotype of US consumers as wedded to the belief in 'free' TV may be breaking down. Will consumers spend an extra $10 a month for information and other services? Will they spend $20 or more? An exact amount is impossible to determine at this point. However, the pay-TV example, the growth of home video games, and the widespread consumer interest in such home-security devices as smoke alarms suggest that there is a potential group of services that consumers will pay for.

The USA may indeed prove to be the slow turtle among the countries exploring the utilities of videotex and teletext. Yet, in the long run, its position could prove healthy for everyone. Because national pride is not vested in any particular system, free and open debate on a wide range of options is possible. Whether the debate will occur, however, is questionable. With a few notable exceptions, past experience shows the USA entering intense debates on the public interest and social benefits of a new technology only after the technology has been chosen. At present, it would be naïvely optimistic to say that the USA is comprehensively reviewing the options and implications of teletext and videotex. Rather, the country has backed into a do-nothing position regarding these new media because the climate makes it difficult and risky to do anything. Whether this passive malaise will be turned into active exploration of problems and possibilities is a basic question at the beginning of the 1980s.

21. Canada

Martin Lane

For videotex in Canada, 1980/81 will be the time of the field trials when a variety of different approaches to home information and security systems will be assessed in terms of their technical suitability and reliability. In terms of timing, then, Canada can be said to be two years behind the front-runner in videotex—Prestel. However, this may be a very misleading assessment. The technology that is being used in the videotex trials in Canada can be said to be at least two years ahead of any of the European videotex systems and it seems likely that if the field trials are successful, many will develop rapidly into public service without the formality of a market trial. It is quite possible that by 1982, videotex will be well into a public service in several provinces. Canada is unique in many ways among those countries currently involved with videotex—it is unique in the technology it has adopted, in its telecommunications organization, and in the information environment into which videotex will be introduced. This chapter looks at developments and attempts to assess the impact of these unique factors on the way that videotex may evolve.

The technology

Canada is the first country to develop and implement what can truly be described as a second-generation videotex system. This system, named Telidon, can immediately be seen to be different from existing systems owing to the better quality of its graphic display. However, Telidon is different from first-generation videotex systems in more than just its graphics capabilities; it is based on an entirely different conceptual approach to the composition and display of a videotex page.

Previous videotex systems have used the alpha-mosaic system for encoding and displaying pages. For instance, on Prestel each page is divided into a mosaic of 40×24 for text and 80×72 for graphics, and the page is composed of letters of text (or blocks of colour in the case of graphics) occupying particular spaces on particular lines within that mosaic. This means that regardless of the resolution of the videotex display unit or the intelligence in the user terminal, the page will always look the same. It also means that since the computer must keep track of each line and space of each page, a considerable amount of memory is required to store each videotex page. Telidon, however, is an alpha-geometric system. Pages are encoded as picture description instructions in terms of standard graphical language. That is to say, the contents of a page are 'remembered' in terms of lines, arcs, polygons, rectangles, and points, with text being a special type of graphic image. For example, a page might consist of: a dark blue rectangle, diagonal

220

from point 1 to point 600; a red circle, radius 20 points, centre at point 400; text string 'Infomart' starting at point 200; etc.

The advantage of this method of storing pages is that it takes less computer memory and less transmission time per page and, more crucially, the page description is independent of the terminal, so that the same page can be seen on a variety of terminals with different resolutions and characteristics and the page is always displayed to the maximum capabilities of the particular terminal. In contrast, in order for systems such as Prestel to take advantage of cheaper memory and better terminals by upgrading its display characteristics, a complete rewrite of the existing software would be required.

The disadvantage of Telidon, at least in the short term, is that more intelligence is required in the user terminal/decoder, which makes it more expensive. Of course, as the price of this intelligence continues to decline this will be less and less of a problem. The commonly quoted mass-production price for a Telidon decoder is $200 (Can.) compared with around $1000 at the moment. As elsewhere, the problem of bridging the gap between the hand-made and the mass-production phase has yet to be solved.

The industry

The Canadian videotex industry, though embryonic, can be seen to be developing along slightly different lines from that in Europe. This is largely because most of the telephone companies in Canada are not becoming involved in the storage and processing aspects of videotex; instead they are sticking to their traditional roles as common-carriers. This means that in the place of the usual videotex triangle of terminal manufacturer–telephone company–information provider, the Canadian industry is more of a rectangle with service companies, offering storage and processing, as the additional element.

As might be expected, there is a considerable blurring of the dividing lines between these four elements, so that some information providers, such as Infomart, are also service companies, and some telephone companies are both information providers and service companies. Additionally, cable TV operators are anxious to participate in interactive services and they see themselves as carriers, service companies, and information providers, and perhaps also as terminal distributors.

Looking at each of the industry elements in a little more depth:

The terminal manufacturers. There are four companies manufacturing subscriber terminals—by which is meant 'black box' decoders that can be plugged into a normal TV—for the Canadian field trials. All of them are in the electronics/telecommunications business and none of them are TV manufacturers. About 1500 terminals will have been built by the end of the year.

The common carriers. These are principally the telephone companies but there are also some cable TV companies involved. Both see the development of videotex as a new source of revenue. Perhaps it would be useful at this point to say that the telecommunications set-up in Canada is very confused and confusing—some telephone companies are provincially regulated and some are federally regulated; some control cable TV networks; while others treat the cable TV industry as a mortal enemy. So there is a situation in Ontario/Quebec where Bell, the telephone company for the two provinces, is totally opposed to any involvement by the cable companies, while in Manitoba, the Manitoba Telephone System is running its videotex field trial via a cable network. Additionally, each telephone company seems intent on demonstrating its superiority over its rivals by developing a slightly different version of Telidon so that compatability between the different field trials in Canada is still an unresolved issue.

The service companies. These companies will provide the storage and processing facilities as well as, in some cases, creating, maintaining, and operating databases. Because these companies will be developing software expertise of their own, it is to be expected that the rate of system improvement and augmentation will be faster in Canada than in Europe and that there will be more of an emphasis of turnkey systems for special applications.

The information providers. As in all videotex systems, these are the vital ingredient, and as elsewhere Canadian information providers consist of a cross-section of publishers, retailers, travel companies, government and educational organisations, specialist information companies, and videotex-only entrepeneurial companies.

The field trials

There are three field trials under way in Canada with more pending. The three actually in operation—all residential trials—are:

1. *Project IDA* (named afted Ida Cates, the first female telephone operator in Manitoba). Run by the Manitoba Telephone System (MTS) in Winnipeg, IDA is testing a broad-band cable system with embedded microprocessors. One hundred homes have been wired to test such things as meter reading, fire and burglary alarms, and thirty of these homes will be supplied with Telidon terminals. In addition to the Telidon trial, there will also be twenty homes with terminals linked to another version of videotex—Omnitex. This is a system broadly similar to alpha-mosaic videotex but with the added facility of keyword searching, with all user terminals equipped with full alphanumeric keyboards.

    For this trial, the MTS is acting purely as a common-carrier and information providers wishing to participate have to either provide their own computer hardware and software or participate via a service company. About 10 000 pages will be available on IDA. If the field trials prove successful, it is generally believed that the MTS plans to expand into a market trial in very short order with the ultimate objective of wiring the entire city of Winnipeg (population 600 000).

2. *Profect VIDON.* Run by Alberto Government Telephones (AGT) in Calgary, VIDON uses a second paired wire into the home which in effect means that videotex can be used without tying up the normal telephone line. AGT will be testing a variety of home alarm systems in 300 homes, 115 of

which will also be equipped with Telidon terminals. The Telidon terminals being used in VIDON are a 'stripped-down' version that have graphic display capabilities closer to those of alpha-geometric systems than to a normal Telidon terminal. AGT has made the decision to go with this version of terminal in the belief that its lower cost will make a videotex mass market more quickly realizable. And, owing to the nature of Telidon, it will be possible to up-grade the terminals as memory becomes cheaper without having to redesign the database.

The AGT will be offering some storage and processing facilities for VIDON as well as allowing users to be switched to service companies' machines. About 10 000 pages will be available on VIDON.

3. *Project VISTA.* Run by Bell Canada in Toronto and Montreal, this system uses the normal telephone line. One thousand users will be given terminals for one year. Bell will be providing storage and processing for this trial as well as allowing switching to service companies. If this trial proves successful, it is believed that Bell will move directly into a public service, by-passing the market-trial phase. About 100 000 pages will be available on VISTA.

In addition to these well-advanced trials, New Brunswick has announced initial plans to run a 75-home videotex field trial, BC Tel (owned incidentally by GTE, the US company that is running a Prestel-like videotex trial in the USA) is thought to be planning a trial with an emphasis on business applications, and several cable TV companies are in the advanced planning stages of field or market trials. As far as is known, all of these trials will use Telidon technology.

The Canadian market

The Canadian consumer electronics industry has been traditionally dominated by foreign manufacturers, be it American or Japanese. No television sets are manufactured locally, although some are imported as components to be assembled locally. Since it appears that the 1980s will be dominated by the movement of silicon chips out of the office and into the home, the Canadian government is anxious to get some sort of foot in the door of this potentially enormous industry.

It is in this context that Telidon was developed and is being promoted in Canada. Telidon is seen as potentially a winner both domestically and internationally—particularly in the USA—and this is very important. Since Canada has a population of less than 25 million, it will be difficult for a Telidon industry to do more than survive if it is entirely confined within the country. However, if Telidon can be sold abroad, especially south of the border, then its future will be assured. The Canadian government has, therefore, been exerting considerable pressure both nationally, to make sure that all videotex field trials use Telidon technology, and internationally, to make sure that no videotex standards are adopted that exclude Telidon.

In terms of potential users of videotex, the Canadian market makes a fascinating study. Because of cable TV penetration of over 60 per cent, with the attendant teletext-like news and weather services, and a multitude of radio stations and special-interest publications, the Canadian is probably able to access more information more readily than anyone in the world, including the USA. The interesting question this poses is whether Canada is consequently a good or a bad market for a videotex service.

Does an abundance of readily available information whet or sate the appetite for more information? Strong arguments can be made for either alternative. Some argue that Canadians—at least those in major cities—have so much information available at their fingertips that a videotex service is redundant. However, the balance of evidence seems to support the counter-argument that information begets information and the need for that information. In the business world, at any rate, it is the very proliferation of information— scientific/technical and commercial—that has created the demand for specialist on-line retrieval services providing, in effect, information about information. There is no reason why this same process should not occur in the domestic market-place. Sam Fedida—the father of videotex—has called Prestel 'the first port-of-call for information', which succinctly describes this type of application.

One question that may be answered by the Canadian videotex experience is to what extent graphics are a necessary or useful aspect of videotex. There is no doubt that Telidon looks better than other systems, but will that make any impression on the user who may basically want only to dial up, obtain a very specific piece of information, and then hang up? Another factor in the Canadian market is the rapid evolution of the home computer industry. Some 200 000 pets, apples, etc., are being sold per month in North America. Whether this market is in competition with videotex or provides a market opportunity is not totally clear, but it can be argued that someone who has an intelligent terminal with a keyboard could quite easily be persuaded to buy a little black box that allows his computer to hook into a larger network providing information, transactions, messaging facilities, etc.

Page creation

The question was posed above as to whether the better graphics on Telidon will significantly increase videotex's attraction to the consumer. It is obvious that information providers believe that it will. Almost the first move of any information provider who decides to get into the page-creation business here is to hire a graphics designer. Because Telidon offers greater flexibility and freedom for graphics creation (besides previously mentioned features there are six shades of grey, a choice of textures and fills for the various geometric shapes, eight text sizes, and simple animation effects available), a higher standard of graphic ability is required to create effective pages. Also, because a Telidon page is displayed in the order in which it is created rather than from top left to bottom right, considerable thought has to be given to the impact of the page as it 'unfolds' before the user's eyes. Background must appear before foreground, news headlines before the news story, routing instructions before graphic display, etc.

All this means that page creation for Telidon is more labour-intensive than for earlier systems. And since all page creation is done on stand-alone mini-computer-based editing terminals that cost between $20 000 and $30 000 each, a committed information provider using three terminals is faced with a $100 000 capital investment before he starts. The implication of the high start-up and page-creation costs for IPs on Telidon is that there will be more 'umbrella' IPs in Canada and less man-and-a-dog IPs of the type that were a feature of the early days of Prestel.

Costs and prices

Some mention has been made of the cost of page creation, but what of the user's cost for accessing Telidon? Since for all the field trials currently operating or planned in Canada, videotex terminals and services are being offered free to users, no one, be it carriers, terminal manufacturers or IPs, has had to make the difficult decision on pricing either in terms of magnitude or method. There does seem to be be a consensus that prices will have to be considerably lower than those of Prestel and that a subscription rather than a usage charge is the most attractive method of acquiring users, at least in the early stages of the system. (Although it must be said that some see videotex as the means by which telephone companies wish to introduce Canadians to the idea of paying for local calls by usage as opposed to the current monthly flat-fee method.)

In terms of the costs that the IP will have to recover in order to make a living off Telidon, page creation as previously discussed will be expensive but storage and processing costs should be fairly low due to the fact that there will be several service companies in competition for the business. Most IPs appear concerned to keep costs to the users very low and this probably means that there will be considerable emphasis on promotional/advertising and electronic shopping type of material as the IP attempts to obtain revenue from sources other than the user.

As far as the terminal cost and, just as crucial, availability is concerned, everyone is anxious not to fall into the Prestel trap of having a system working, IPs creating pages, and then finding that there are no user terminals available at anything like a reasonable cost (if at all). However, there is still the problem of convincing the terminal manufacturers that they should invest x million dollars in tooling up a mass-production line before a mass market has been shown to exist. At the moment, all that can be said is that everyone is aware of the problem and that a higher priority is being given to solving it than perhaps has been the case in Europe. As yet though, no magic solution has been found.

Part Six

The future

22. Viewdata and society

Ederyn Williams

It is with some trepidation that I write this chapter. The topic is much too important to ignore; indeed, in some countries, notably Denmark, it is felt necessary to determine the social impact of viewdata before much technical or market development is undertaken. On the other hand, it is very easy to be spectacularly wrong. Predicting the future is always hazardous and nowhere more so than when predicting the use and impact of technical innovations.

Just as it was believed that passenger rail travel was impractical, as the human frame could not survive travelling at over 40 miles per hour, new telecommunications systems have often been felt to be infeasible or unrealistic. In 1879, Sir William Preece, chief engineer of the British Post Office, said of the telephone:

> I fancy the descriptions we get of its use in America are a little exaggerated, though there are conditions in America which necessitate the use of such instruments more than here. Here we have a superabundance of messengers, errand boys and things of that kind. . . . Few have worked at the telephone much more than I. I have one in my office, but more for show. If I want to send a message—I . . . employ a boy to take it. (quoted in Pool, 1977, p. 128).

Not all forecasts have been wildly pessimistic. Some have been wildly optimistic. For example, in 1969 Julius Molnar, Executive Vice President of Bell Telephone Labs, said of Picturephone, their new videotelephone:

> I predict that before the turn of the century Picturephone will . . . displace today's means of communication, and in addition will make many of today's trips unnecessary.

The President of A.T.&T. was even quoted as saying that he expected 1 million Picturephones in the USA by 1980 (*Washington Star,* 1973). In fact, Picturephone has been a real flop: there are probably no more than 1000 in the world today, and growth is virtually zero.

Eminent men have not only been wrong about the amount of use of innovations; predicting the nature of their use is an even more hazardous business. Thomas Edison, inventor of the gramophone, is claimed to have believed that the major use of his new invention would be to record the last words of dying men, so as to avoid disputes between the heirs. Lord Reith, the first Director General of the British Broadcasting Corporation, ventured the opinion, in the early days of radio, that the primary use that he could see for this device was for listening to church sermons, as that was the only occasion he knew when people gathered together just to listen.

Even if one could be certain of the scale and type of use of a new communications medium, one could still very easily miss some extremely important indirect impacts. For example, it has been pointed out (see Pool, 1977, p. 140) that without the telephone, there could not have been skyscrapers as we know them. Skyscrapers have always been a rather marginal economic proposition, though the prestige involved compensates for the slight extra cost. If, however, there were no telephones, there would have to be twice as many lifts to accomodate all the messenger boys, making the whole building thoroughly uneconomic. Yet who could have predicted this impact in the 1880s?

Impact on other media

Just now, the employment impact of microchip electronics is a trendy topic for pundits. Applied to Prestel, the apocalyptic vision is of electronic media replacing newspapers within 10 to 15 years, and bread queues of compositors and printers forming in Fleet Street (some hawks, of course, find this vision quite pleasing). Reality is unlikely to be as dramatic; indeed it seems unlikely that Prestel, however successful, will ever replace newspapers. Old media are rarely completely replaced by the new: either they live on in genteel semi-retirement, or they adapt to a new form, so as to compete less directly. Thus the cinema has not completely replaced live theatre, nor has television completely replaced the cinema: in both cases the older form lives on in reduced circumstances. One 'old' medium, radio, has adapted well to the growth of television: by moving away from networked drama and documentary, towards music and local news, it has gained new energy. Today there are more radio stations, and more work in radio, than before television started.

It is also a mistake to assume that more use of Prestel automatically means less use of newspapers and other printed media. Two other phenomena can be observed. Firstly, each medium can stimulate use of the other; recent examples have shown that people can often be persuaded to buy 'the record of the book of the film of the play'. If printing groups use Prestel energetically, they may find that it increases, rather than reduces, sales of associated printed media. Secondly, new media can often generate new audiences, rather than merely poaching their audience from an established medium. Early results from the Prestel test service illustrate this effect. People who were going to get Prestel in their homes were asked what media they used to obtain various types of information. Two months after they got their Prestel set, they were asked again. Table 22.1 summarizes the shifts.

It is clear that a large proportion of the audience gained by Prestel and teletext had previously not got information from any source. The new media are meeting information needs that the old media were unable to meet—a clear benefit. In comparison, the shifts away from the other media are smaller, and, since they are spread across several different media, seeem unlikely to do great harm to any of them.

My conclusion is that the existing media need not greatly fear Prestel teletext. At best, the old media will merely shift their emphasis so as to compete less directly; at worst, they will lose their licence to print money.

Table 22.1

	Medium	*Percentage shift (%)*
Traffic gained by	Prestel	+ 14.2
	Teletext	+ 8.8
Came a little from	Television or radio	− 2.4
	Telephone contact	− 2.1
	Newspapers	− 1.5
	Libraries	− 1.5
	Journals or magazines	− 1.2
	Books or directories	− 1.0
But mostly from	Nowhere	− 9.3

(N.B. Gains need not balance losses, since people can get their information from more than one source.)

Impact on the information environment

To continue the argument, Prestel, if successful (and I have been assuming that it will be) will enrich the information environment of everyone. The effects that are likely to occur have been neatly summarized by Katzman (1974): I shall entitle them 'the three laws of informatics'.

The first law states: 'All individuals receive more information after the adoption of a new communication technology.' This law suggests that all such media tend to have a beneficial impact, at least until information overload is reached.

The second law introduces a note of concern: 'With the adoption of new communications technology, people who already have high levels of information and ability will gain more than people with lower initial levels. This is an empirical phenomenon which seems to be almost universal.'

The third law, however, indicates that we should not be too concerned: 'There are limits to the amount of information any individual can successfully digest or retain. If information-rich individuals achieve full use of a new communication technology, they cannot further improve their use of that technology. Relatively information-poor groups will then close the gap.'

We can expect these three laws to apply to Prestel. Recognition of the fact that they are universally applicable is important, otherwise we might be misled into accepting that the initial gap-widening impact of Prestel on the information-rich and information-poor means that Prestel is socially a bad thing. True, the implications of the second law are bad, but the first and third laws show that the overall and long-run implications are good. An equality of information-poverty is not a desirable goal.

Impact on individuals

One can speculate at length about the impacts of Prestel on the everyday life of individuals in the 1980s and 1990s. Demonstrating such impacts is more difficult. Early in the television era, there was much detailed research comparing the behaviour and attitudes of people with and without television. A prime example of such work is *Television and the Child,* by Himmelweit, Oppenheim, and Vince (1958), but a reading of the work shows that it was far from easy to show that television had any important effect on behaviour and attitudes at all. The problem of proving effects has not stopped people pontificating on the subject since. The problem may be, despite Marshall McLuhan's beliefs, that the specific effects of a medium are few compared to the effects of content: the portrayal of violence may have much the same impact whether in books, comics, cinema, radio, or television.

I would thus expect the effects of Prestel on individuals to be minor rather than major, and gradual rather than sudden. I do not expect it to be a force lifting the masses from the mire of trivia that they normally read: Prestel will tend to have just as much trivia as other media, no more, no less, because it is that amount of trivia that people wish to read. Prestel may have beneficial effects in informing the ignorant and educating the young, but it is unlikely to achieve goals that could not have been reached through the printed word. I would, however, venture to suggest two effects on individuals that could be fairly specifically attributed to Prestel, or at least to viewdata-like media.

Firstly, Prestel is an active medium, in that nothing happens unless the user presses the buttons. The users of Prestel are not viewers or watchers, nor can they really be called an audience, as all these terms are appropriate to media like television, film, or radio where the user's role is to passively let the flow of material wash over him. The Prestel user is more of an operator or reader, who actively chooses every second whether to move on, and what to look at next. Incidentally, this active/passive classification of media bears some similarity to McLuhan's division of media into the 'hot' media (where much is given and little has to be filled in by the user) and 'cool' media (which give little and require much to be filled in). However, I would consider film and television as equally passive, requiring little from the user, while McLuhan calls film hot, like radio and the printed word, but TV cool, like the telephone. I adopt my own distinction because I find his unhelpful and ultimately confusing.

Since Prestel is an active medium, requiring activity from its users, it should have quite different impacts from broadcast television. Most of the readers who have children will recollect having said to them, at one time or another, 'why don't you switch off the television and read a book'. Even though the television may be showing *Hamlet,* while the book is just *Batman,* many of us seem to feel that the book is more acceptable, more educational. In part, this must be mere traditionalism, but the fact that reading is an active pursuit, requiring knowledge, learning, thought, and movement is also an important part of our attitude. The objection to television is due not just to the quality of some of the programmes, but also to the passive, vegetable state that it seems to induce in viewers. Prestel, however, is clearly an active medium. I predict, therefore, that it will be widely acceptable as a quasi-educational tool, rather like home encyclopaedias. People who object to watching a lot of TV, indeed, people who refuse to own one, will see the incorporation of the Prestel facility as a saving grace. Prestel will be much more readily accepted as 'good for you', and its combination in the same device as broadcast television, which is generally thought of as 'bad for you', will create some interesting tensions.

The requirement that Prestel puts upon the user to actively control his television set, his telephone, and the Prestel computer in order to get what he wants will generate an interesting result—the second specific effect of Prestel. Prestel, as the simplest of a family of on-line computing systems, will act as an ideal introduction to more complex systems, and will introduce its users into the complex world of on-line computing, micro-computers, word processors, and all the other devices that we need to learn to live with. Just as school essays train journalists, and school maths trains accountants, so Prestel, especially in the home, will train the generation of children already born for whom almost any future job will involve control of computer systems or other intelligent machines. Many computer experts have expressed surprise at the relatively slow rate of penetration of computers into business; the ultimate impediment always seems to be the people who must control, or at least understand, the computers in order to turn them to profitable use. Since Prestel is so simple that any child who can read can learn to operate it, and is in addition sufficiently interesting and entertaining to motivate them to learn, it is likely to have a major role in teaching the next generation to use intelligent machines in a more adept and appropriate fashion than their fathers are capable of. This could be the major educational benefit of Prestel.

Impact on professional middlemen

Prestel is a means for direct information transmission into the home and business. Inevitably, it will alter the flows of information. Particularly, its power in rapid and flexible information transmission may make it easier for the end consumer of information to short-cut the usual flow, avoiding the middlemen who previously retransmitted, with or without repackaging, information from the original source. This could make life tough for professional middlemen, among whom one must count journalists, teachers, travel agents, insurance brokers, estate agents, and information scientists. Ivan Illich in various writings, for example his *Tools for Conviviality* (1973), is especially scathing about the role of the professional intermediary. He advances the theory that individual satisfaction is related not merely to the material goods owned, but to the opportunity given to people to creatively manipulate their environment. Tools (by which he means any mechanical devices) which are simple to use and available to everyman, he calls convivial tools. According to Illich only such convivial tools are ultimately beneficial to society and the individuals in it. To him, many of the present ills of our society are created by unconvivial tools—tools which cannot be used by everyman, either because they are intrinsically too complex, or because their use has been cornered by an exclusive group of professionals. The telephone is clearly a convivial tool: anyone can use it for any conversational purpose he wishes, as and when he wants to. The computer, on the other hand, has been almost the ultimate in unconviviality: its complexity

233

defies the puny efforts of everyman and the machine becomes the property of a select priesthood of computer analysts and programmers, leaving everyman to curse the machine from outside.

The beauty of Prestel is that it is the first convivial computer system: available 24 hours a day to anyone who can read (and can afford it), it strips away the mystique of using computers, and of manipulating information in one blow. With a convivial tool like this put in the hands of everyman, what is to stop him becoming more and more independent in his habits, gradually leaning less and less on the middlemen? Such a trend certainly seems possible, and it fits the popular ethos of self-service and self-sufficiency.

Should the professional intermediaries be quaking in their boots? Certainly if they are complacent, convinced of their own worth, while ignoring developments in the electronic information media, then they could be swept away within a generation. However, the future is not really that bleak for the middlemen. Most people are suffering from some kind of overload on their information-processing capabilities. Although sophisticated international data network might make it possible for an individual to make up his own holiday, booking his own room direct with the hotel, his own airplane ticket, his own hire car, and coach ticket, most people will find this too arduous, and will take the package holiday instead. As long as the intermediaries are processing the information in expert ways, then they are actually adding value, and need not fear replacement by electronic information media. Doctors need hardly fear that putting a medical encyclopaedia on Prestel would lead to patients going in for extensive self-diagnosis, because the doctor is (or should be) using years of expertise to arrive at a diagnosis and treatment in a way that could not be duplicated by a layman, whatever the information sources available to him. However, other, less expert, intermediaries may need to be more thoughtful about their future role.

Fragmentation of the mass culture

At present, we live in a mass culture, where most people tend to read the same news and watch the same TV programmes and thus tend to have similar knowledge, interests, and attitudes. This tendancy is, at least in part, the product of the mass media: each new medium, from newspapers, through cinema and radio, to television, has tended to be highly monolithic and centralized—produced by the few for the many millions.

Clearly the mass media have some positive benefits, in that they strengthen the consensus, and the major interests of the society, increasing social cohesion and diminishing the extent to which misunderstandings between people can arise merely because of disparate experiences producing different language, behaviour, and modes of thought. In every revolution, the first aim of the revolutionaries is to capture the presses and the broadcasting station, as they know that these are the kingpins that dominate society.

However, it has been persuasively argued by Maisel (1973) in a paper entitled 'Decline of the mass media', that the mass media, such as networked television, national newspapers, and general-interest magazines are already in decline.

Certainly both in Europe and the USA, there is evidence that such media are not merely witnessing a declining audience share, but in some cases are shrinking in absolute terms. This decline, according to Maisel, is not only inevitable, but is probably desirable. The mass media were but the first stage of mass production: just as the Model T Ford came in any colour as long as it was black, but modern Fords come in thousands of shapes, sizes, and colours, so the mass media are giving way to more varied, individual forms. More specialized media, such as local radio, cable TV, specialist magazines from *Gay News* to *Hot Car*, and local newspapers are all enjoying explosive growth.

Prestel is well placed to take advantage of, and to reinforce, this trend. Hooper (1979) has dubbed it 'the first individualized mass medium', but in fact its only mass-medium characteristic is the number of customers it is aiming to attract. From the user's viewpoint, it is highly specialized, indeed individualized. Each user reads just the pages he wants from the hundreds of thousands available: if he is interested only in chess, dog breeding, and African politics, he can read just that material, without being even momentarily distracted by, or paying for, other material. (Incidentally, it is a common fallacy among newspaper men and others wedded to the printed word, to calculate how much it would cost a user to read all the material in a daily newspaper if it was on Prestel. The comparison is false, as very few readers read everything in the newspaper, and, if they do not, then they are paying for unused information. On Prestel, they need only pay for what they actually use, so discriminating readers can get a much better bargain.)

Prestel, then, is offering not merely a specialized service, like a special-interest magazine; it is offering page-sized chunks of information and entertainment, that can be combined in any way the user wishes to make his own individualized magazine. Future developments in the system may make this process even more automatic: perhaps a user could choose to have chess, and only chess news automatically sent to his terminal from the Prestel computer as and when such news was available. All other news, no matter how important, would not then be brought to his attention.

Prestel, then, will reinforce the trend away from the mass media and mass culture towards the specialized and individualized. Whether one sees it as good or bad depends on one's viewpoint: certainly it makes it more difficult to wield power and influence from the centre, but the richness and variety of a plural society should be enough to compensate for the lessening of cohesion and consensus.

The home office of the future

More speculatively, one can suppose that Prestel, together with other telecommunications advances like teleconferencing, fast facsimile and cheap word processing will diminish the distinction between home and office, and lessen the need for office work to be concentrated in the centres of large cities. Meier (1962) has asked, 'Why must almost all major policy makers congregate with their assistants and ancillary help in the same vicinity at the same time? It has been observed that they spend their time there manipulating symbols which are directed at others. Why not transmit the symbols over a somewhat greater distance instead of moving people? Isn't this the ultimate solution' (p 60).

Increasingly, people are finding it possible to work much of their time at home or at a small local office, only occasionally needing to venture into the big city for especially important meetings. Prestel will make this easier for them, as a ready flow of accurate, up-to-date information has in the past been difficult in geographically outlying areas. Together with other telecommunications and computing developments, it could make it much easier for a larger proportion of people doing white-collar work to be in the most convenient, or most pleasant locations, rather than being required to commute by overloaded transport systems to an overcrowded city centre. Some people might find the social stresses of working from their own home too great, and would prefer an office just around the corner. With the wide range of modern telecommunications systems now being developed, the choice may be theirs before the end of the century. The benefits and problems of such a shift away from the big city have been described elsewhere (e.g., Goldmark, 1973): suffice it to say that Prestel is one element of the electronics revolution necessary to give us the choice.

Conclusions

Of the various possible impacts, each has positive and negative elements, and in many cases how one judges these effects depends on one's ideological leanings. Our society may be changed by Prestel, but it would be wrong to assume that the changes will be all good or all bad. To quote Reid (1978): 'the electronics revolution should be treated as innocent until proved guilty. To look on the bright side, Prestel does not kill, pollute or make your teeth fall out.' We must not ignore the social consequences of our innovation, but my belief is that the likelihood of such impacts being, on balance, detrimental to our society or the individuals in it, is sufficiently small that we need not slow technical or marketing progress merely because we are not yet omniscient.

References

Goldmark, P. C. (1973) 'Tomorrow we will communicate to our jobs', *The Futurist*, pp. 55–58.

Himmelweit, H., Oppenheim, A. N., and Vince, P. (1958) *Television and the Child: An Empirical Study of the Effect of Television on the Young*, Oxford University Press, London.

Hooper, R. (1979) *Prestel: The Editorial Opportunity*, Mills and Allen Communications, London.

Illich, I. D. (1973) *Tools for Conviviality*, Calder and Boyars, London.

Katzman, N. (1974) 'The impact of communications technology: Promises and prospects', *Journal of Communications*, **24**, 47–59.

McLuhan, M. (1964) *Understanding Media*, Routledge and Kegan Paul, London.

Maisel, R. (1973) 'Decline of the Mass Media', *Public Opinion Quarterly*, **37**, 159–170.

Meier, R. L. (1962) *A Communications Theory of Urban Growth,* MIT Press, Boston Mass.

Molnar, J. P. (1969) 'Picturephone service – A new way of communicating', *Bell Laboratories Record,* **47,** 134–135.

Pool, I. de S. (1977) *The Social Impact of the Telephone,* MIT Press, Boston Mass.

Reid, A. A. L. (1978) 'The sky's the limit', *The Prestel User's Guide,* **1,** 3.

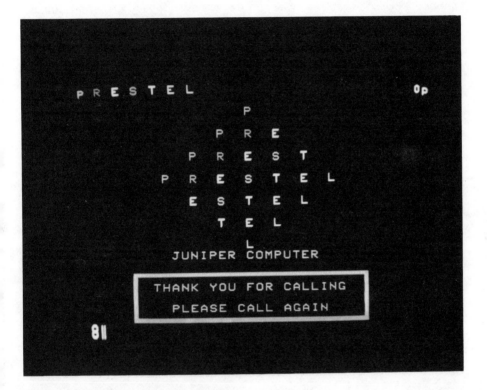